VISUAL RHETORIC SERIES
EDITED BY MARGUERITE HELMERS

LOCATING VISUAL-MATERIAL RHETORICS

The Map, the Mill, and the GPS

Amy D. Propen

Parlor Press
Anderson, South Carolina
www.parlorpress.com

Parlor Press LLC, Anderson, South Carolina, USA

SAN: 254-8879

Library of Congress Cataloging-in-Publication Data

Propen, Amy D.
 Locating visual-material rhetorics : the map, the mill, and the GPS / Amy
D. Propen.
     p. cm. -- (Visual rhetoric)
 Includes bibliographical references and index.
 ISBN 978-1-60235-254-4 (pbk. : alk. paper) -- ISBN 978-1-60235-255-1
(hardcover : alk. paper) -- ISBN 978-1-60235-256-8 (adobe ebook : alk.
paper) -- ISBN 978-1-60235-257-5 (epub : alk. paper)
 1. Visual communication. 2. Visual perception. 3. Space perception.
4. Material culture. 5. Cartography--Social aspects. 6. Public spaces-
-Social aspects. 7. Geographic information systems. I. Title.
 P93.5.P76 2012
 302.2'22--dc23
                              2011036238

Cover design by David Blakesley.
Cover image: "Topographic Map." © 2006 by Brandon Laufenberg. Used
     by permission.
Printed on acid-free paper.

Parlor Press, LLC is an independent publisher of scholarly and trade titles
in print and multimedia formats. This book is available in paper, cloth and
eBook formats from Parlor Press on the World Wide Web at http://www.
parlorpress.com or through online and brick-and-mortar bookstores. For
submission information or to find out about Parlor Press publications, write
to Parlor Press, 3015 Brackenberry Drive, Anderson, South Carolina, 29621,
or email editor@parlorpress.com.

Contents

Illustrations

Acknowledgments

I would like to thank the many people who have provided me with support, assistance, and guidance during the writing of this book, starting with my new friends and colleagues at the Lowell Mills National Historical Park. Tess Shatzer was hugely helpful and patient in providing detailed information about the history of both the park and the Mill Girls. Andy Pearson's insightful and detailed boat tour of the Lowell canal system informed my understanding of Lowell's geographic history and its relationship to the city's industrial heritage. Thanks to Jack Herlihy at the Division of Cultural Resources for his prompt responses to my abundant inquires about documenting park resources and artifacts. Thanks also to Martha Mayo at the Center for Lowell History for helping to clarify information about some of the early writings of the Mill Girls. The assistance and guidance of these individuals informed my understanding of the mills and my writing of chapter three.

I also wish to thank Martin Dodge, whose helpful suggestions related to the study of GPS use pointed me in some fruitful directions that eventually led to the research and writing of chapter four. My engaging conversations with Krista Kennedy about the interactiveness of agency also informed my rhetorical analyses of GPS use. And of course, I am most grateful to the twenty-two people who shared with me their GPS stories. They spoke with great seriousness, candor, and often good humor about their interactions with the GPS, and I have done my best to preserve and convey their stories here. This book would not be possible without them.

Thank you to Joel Reynolds at the Natural Resources Defense Council for permission to reprint the map in Figure 46, and to Joep Luijten for his input, expertise, and feedback during the early writing and revisions of chapter five. Francis Harvey also helped me understand the societal dimensions of cartographic practice in more nuanced ways that helped shape the early writing and subsequent revisions of chapter five.

Portions of this book have appeared in previous form and are reprinted with permission. Chapter five was reprinted with modification from "Visual Communication and the Map: How Maps as Visual Objects Convey Meaning in Specific Contexts," published in *Technical Communication Quarterly* 16.2 (2007). Portions of chapter one were reprinted with modification from "Cartographic Representation and the Construction of Lived Worlds: Understanding Cartographic Practice as Embodied Knowledge," which originally appeared in *Rethinking Maps: New Frontiers in Cartographic Theory,* edited by Martin Dodge, Rob Kitchin, and Chris Perkins (New York: Routledge Studies in Human Geography, 2009). I gratefully acknowledge permission from the Taylor and Francis Group to reproduce modified versions of both of these works here.

Sections of this book are also based on what was originally my dissertation project. Thus this book would not have come to fruition without the support of my committee at the University of Minnesota. In particular, this project would not have been possible, in any of its iterations, without the unwavering support and direction of Mary Lay Schuster. It was Mary who first introduced me to the writings of the Mill Girls and encouraged me to tell their story. Her insights and perspective allowed me to understand material rhetorics in ways that I would not have otherwise considered and helped move my thinking in new directions. I owe her much gratitude, not only as a dissertation director and mentor but as a friend and colleague. Richard Graff's insights, particularly related to the presence of rhetorical figures in the Lowell park artifacts, helped further my analysis in chapter three. Laura Gurak continues to provide invaluable suggestions and resources related to digital rhetorics, not to mention good advice related to the sport of kayaking. Art Walzer provided valuable opportunities for the discussion of Foucault's theory of discourse and continues to be a multimodal source of support and encouragement. Michael Salvo and Elizabeth Shea also helped guide my initial thinking on the intersections of visual rhetoric and critical cartography. I would also like to thank Beth Britt at Northeastern University, not only for first introducing me to material rhetorics and the work of Carole Blair but also for her continued support and mentoring over the years.

I will always recall with fondness and perhaps a dangerous nostalgia the many conversations and memorable experiences fostered through the supportive environment of the Rhetoric Department on the St. Paul campus. Friends and colleagues too numerous to name contribute to these memories, though I would especially like to thank Paul Anheier, Kenny Fountain, Cristina Hanganu-Bresch, Clancy Ratliff, and Greg Schneider both for those memorable moments and for the sustained connections. Thanks also to Jessica Reyman who, on more than one occasion, talked me through my writing angst, long distance, from Illinois to Pennsylvania.

A summer research grant and subsequent fall course release provided by York College of Pennsylvania greatly facilitated the writing of this book. In addition, I would like to thank my colleagues in the Department of English and Humanities for their support, ideas, and encouragement, including Gabriel Abudu, Dominic DelliCarpini, Madeline Yonker, and Mike Zerbe.

Several colleagues provided valuable insights and ideas in the shaping of this project. Mary Lay Schuster kindly provided early feedback on the proposal for this book. Dave Blakesley at Parlor Press and series editor Marguerite Helmers reviewed the proposal with enthusiasm and offered guidance and important suggestions. Marguerite graciously provided an early manuscript review that helped me to further conceptualize the book's theoretical work. Both Dave and Marguerite provided great support in preparation and revision of the manuscript and answered my many questions along the way. Madeleine Sorapure at the University of California Santa Barbara provided a highly useful manuscript review; her suggestions allowed me to better contextualize my analyses and forge a clear path for the reader. I am also very grateful for the careful and constructive copyediting performed by Kristen Seas Trader.

I would like to thank my family: Beverly, Michael, Mindy, and David Propen, for their continued support throughout the years. My grandfather, Fred Schoen, and my aunt, Doris Pritt, are not here to see this project completed, but their generosity and unconditional support continue to be a source of motivation and inspiration for me.

Finally, I thank my partner, Karen Dias, whose perspective, encouragement, and unique capacity for patience ultimately made this book possible.

Introduction

Every year, nearly 700,000 people visit the Lowell Mills National Historical Park in Lowell, Massachusetts. Home to the historic mills of the early New England textile industry, the park tells the story of the industrial revolution through its restored mills and boarding houses, exhibits, parks, green spaces, and public art installations (*Lowell National Historical Park*). When describing visitors' first reactions upon entering the park and seeing the historic textile mills where female workers labored in the 1800s, a long-time park ranger there notes without hesitation that visitors "are struck by the size of the buildings." Even those visitors who come to the park with knowledge of the textile industry in early New England, or who have ancestors who worked at the Lowell Mills, the ranger says, "see the high ceilings [. . .] they see the brick [. . .] the physicality of the buildings," and experience the site differently than they have through texts. Visitors are struck by the fact that the mill workers lived next to the huge factories where they labored for fourteen hours a day. Moreover, when visitors juxtapose the size of the boardinghouses with that of the Agent's House, they begin to make connections about the unfairness of the workers' living conditions; that is, the Agent's House (now the Park Headquarters) was home to the agent's family, "but is about the same size as a boardinghouse, which housed up to 250 women" (Park Ranger). Thus the visual and cultural landscape of the park "drives home the tension between emerging classes" (Park Ranger). As the ranger describes, many visitors soon note that they "don't feel like they're getting the big picture" of the history of the Lowell Mills unless they make the time to "see everything." For visitors to the Lowell Mills National Historical Park, then, "seeing comes before words" (Berger 7).

When, in 1972, the art historian and critic John Berger first wrote that now-familiar sentence, "seeing comes before words," he might not have guessed that his influential essay, "Ways of Seeing," would be invoked to support a park ranger's ideas about how visitors' initial experiences of an historic site are visual and corporeal rather than expressly verbal. Nor might he have guessed that his essay would later be anthologized in collections such as Bartholomae and Petrosky's *Ways of Reading* and subsequently become the inspiration for many a first-year writing assignment, or that his ideas would be used to help describe the subdiscipline of visual rhetoric at the beginning of a book about the rhetorical elements of national parks, maps used in environmental debates, and in-car navigational devices like the global positioning system (GPS). Nonetheless, Berger's ideas about how we see in many ways constitute what we might refer to as visual culture and can serve as an accessible point of entry for understanding what, in the mid-1990s, became known among scholars of rhetoric and composition as the burgeoning subdiscipline of visual rhetoric.[1] "Ways of Seeing" is perhaps most well-known, at least among those who have taught the essay in their first-year composition courses, for its ability to prompt discussions about how we see based on what we know. That is, our prior knowledge, cultural contexts, and learned assumptions about the world around us influence our interpretations of visual artifacts like, as Berger argues, paintings and photographs and, as this book will soon discuss, physical sites and material artifacts such as parks, green spaces, and public monuments. In the well-known quotation that opens the essay, Berger emphasizes the prevalence of the visual within society when writes that "seeing comes before words" (7). Rather than create a binary between word and image, he sees an ongoing interplay between them: "It is seeing which establishes our place in the surrounding world; we explain that world with words, but words can never undo the fact that we are surrounded by it. The relation between what we see and what we know is never settled" (7). While this book starts from the point of assuming a wider array of devices than just words to help "explain the world"—again, public monuments, cartographic representations, and even multimodal devices like the GPS are all artifacts that help us interpret and explain the world and, as we will see, are themselves products of visual culture—Berger's larger point is well-received and holds true today: the relationship between what we see and what we know is always shifting and is a product of changing cultural contexts, public understanding, and modes of human communication. Thus, what Berger alludes to here is in fact a working definition of what may be understood as *visual culture*.

As visual rhetoric scholar Cara A. Finnegan notes, there are many ways to understand the concept of visual culture, but broadly construed, it "recognizes that visuality frames our experience and acknowledges 'that vision is a mode of cultural expression and human communication as fundamental and widespread as language'" ("Recognizing Lincoln" 62). *Visual rhetoric* then draws on visual culture to consider the ways in which rhetorical action is "enacted primarily through visual means, made meaningful through culturally derived ways of looking and seeing and endeavoring to influence diverse publics" (Olson et al. 3). Visual rhetoric is likewise attuned to the many persuasive components of visual artifacts and how they function relative to specific audiences, or the social contexts that shape how such artifacts might be interpreted by their viewer. Berger's focus in "Ways of Seeing," for example, is largely on the assumptions that viewers bring to their interpretation of a given work of art. Through his analysis of two portraits created by the painter Frans Hals, Berger argues that works of art can serve to obscure or revise history, as viewers bring their own learned assumptions to bear on interpretations of these visual objects. When we interpret works of art or other visual artifacts based on our own learned assumptions about "beauty, truth, genius, civilization, form, status, taste, etc.," he writes, we perpetuate what he calls an obscuring or "mystification" of the image, one that may work to distance the viewer from the artifact's original meaning or context (Berger 11). Moreover, images are invariably reproduced over time (for example, in advertisements or photographs, or in sculptures or on websites). On the one hand, reproductions and appropriations make famous works of art accessible to the public. On the other hand, because those reproductions tend to manifest mostly in advertising images and in the mass media, they not only perpetuate capitalism, as Richards and David suggest in their discussion of "Ways of Seeing," but also create a sort of mystification that distances viewers from the work's original context and meaning (7). As social understandings continue to change, our interpretations of that which was originally represented by the image will likewise continue to shift.

The shifting interpretations that these visual artifacts can help perpetuate, then, have varied consequences. For, as Robert Hariman and John Louis Lucaites note, popular media and the arts (specifically photojournalism, as they discuss) "can extend an essential but imperfect capacity for connecting with and caring for others"; to do so, however, "they have to be capable of being misleading or misused" (92). As Hariman and Lucaites describe of iconic photographs, and as I also suggest can be the case, though to different ends, with visual and material artifacts such as cartographic representations, green spaces, and public monuments,

> [t]hey provide models for action and assurances that we need not lose what we value most. Ultimately, they function as evidence of things unseen, referring not just to what has past but to what always is outside of our given frame of perception. Whether such images will serve the ends of mystification [. .] or movement towards a better life they cannot themselves represent, remains to be seen. (92)

As I hope to show in this book, our perceptions of visual and material artifacts and the interpretations that such artifacts help foster can have varied consequences not only on our understandings of history but also on our individual, lived experiences and for broader societal issues such as legislation and policy-making. To understand visual artifacts (like photographs and maps) or physical sites (like green spaces and public monuments) as able to shape our understanding of the world around us means understanding these artifacts as rhetorical, or as Carole Blair has put it, as "partisan, meaningful, and influential," to the extent that they have the capacity for consequence, have the ability to persuade, and may influence our interpretations and understandings of specific contexts in ways that impact both the mind and body (Blair and Michel, "Commemorating" 72). Rhetorical criticism is typically concerned with the study of text and discourse in order to achieve "a greater understanding of human action" (Segal 2). To account also for the visual and material within rhetorical criticism then involves two main components: first, as Finnegan argues in "Doing Rhetorical History of the Visual," we must understand the visual and textual and, as this book will soon argue, the material, not from the point of their distinction but from the point of their interplay; second, we must understand visual and, again, material rhetorics "as something more than merely a genre category or product." That is, on the one hand, a photograph or map would count as an artifact of visual rhetoric because "it consists of non-textual or non-discursive features." On the other hand, to understand the photo or map in this way not only serves to perpetuate a visual-verbal divide, but may also be viewed as subordinating *visual rhetoric* to broader studies of text and discourse, which then get to count as "just *rhetoric*." To account for a more inclusive understanding of the artifacts of rhetorical criticism, Finnegan suggests that we "conceptualize visual rhetoric as a mode of inquiry, defined as a critical and theoretical orientation that makes issues of visuality relevant to rhetorical theory." As such, she says, the "visual rhetoric project would urge us to explore our understandings of visual culture in light of the questions of rhetorical theory, and at the same time encourage us to (re)consider aspects of rhetorical theory" relative to the new challenges brought about through analyses of visual artifacts. Projects of visual rhetoric would then understand visual culture as able "to illuminate the complex dynamics of power and knowledge at play in and around images"; they would also understand the "complexities of the relationships between images and texts" as opening up rather than closing off interpretive possibilities (Finnegan, "Rhetorical History" 198). As I will soon discuss in more depth, it is a task of this book to show how material and multimodal rhetorical artifacts are also implicated in the projects of visual rhetoric, and subsequently, to illuminate a more inclusive understanding of the projects of visual rhetoric through what I will call visual-material rhetorical analysis.

Before moving forward, however, it is necessary to better explain what we mean by "the projects of visual rhetoric" in the first place, or how we might apply an understanding of visual rhetoric as a mode or project of inquiry. To do so, I highlight two such projects stemming from the recent work of Finnegan and Hariman and Lucaites, who focus primarily on the visual genre of photography, arguably one of the genres most readily associated with studies of visual rhetoric.[2] I should note that my aim here is not necessarily to provide an extensive review of scholarship in visual rhetoric, nor is it to paint an overly narrow picture of what studies of visual rhetoric ought to resemble. Rather, I am interested in describing for the reader who is perhaps less familiar with the subdiscipline some clear ideas about what approaches to visual rhetoric might entail or what it might mean to understand visual rhetoric as a mode of inquiry. Namely, Finnegan's study of Abraham Lincoln and what she calls *image vernaculars,* and Hariman and Lucaites's work with iconic photographs, engage nicely the components necessary for understanding visual rhetoric as more than a product or mere genre category. Thus, a general understanding of the goals of their work and what visual rhetoric projects can "look" like will help provide a more solid foundation or schema for discussing visual rhetoric, thereby allowing the reader to build on that understanding when, following these initial discussions, I will situate visual rhetoric more specifically in terms of its relationship to studies of space, place, and cartography and describe its more material and embodied components.

Using the tools of rhetoric and informed by understandings of visual culture, Finnegan analyzes the earliest known photograph of Abraham Lincoln, a daguerreotype that dates to the 1840s and was later published in *McClure's* magazine in 1895. To carry out her analysis, she says, "requires careful, situated investigation of the social, cultural, and political work that visual communication is meant to do" ("Recognizing Lincoln" 62). She situates the rare photo of an uncharacteristically well-coifed, youthful-looking, head-to-shoulders portrait of Lincoln in what she calls the "*image vernaculars* of late nineteenth-century visual culture" ("Recognizing" 62). Doing so allows her to fulfill the three main criteria of a visual rhetoric project. First, by understanding image vernaculars as "enthymematic modes of reasoning employed by audiences in the context of specific practices of reading and viewing in visual cultures" ("Recognizing Lincoln" 62–3), she is able to understand the artifacts of visual culture "in light of the questions of rhetorical theory," while simultaneously situating rhetorical theory relative to analyses of visual artifacts ("Rhetorical History" 198).[3] This approach then paves the way for fulfillment of the subsequent criteria of the visual rhetoric project. That is, next, Finnegan's analytical approach "illuminate[s] the complex dynamics of power and knowledge at play in and around images" ("Rhetorical History" 198) by revealing that readers' overwhelming responses to the daguerreotype reproduction in *McClure's* "tapped into myths about Lincoln circulating in the late nineteenth century" ("Recognizing" 62). Based on their understandings of photography at the time and interpretations of the "'scientific' discourses of character such as physiognomy and phrenology," readers felt comfortable analyzing the physical qualities of the Lincoln they saw in the photo ("Recognizing" 62). They then recognized these traits as evidence of his moral character and used the photo "to elaborate an Anglo-Saxon national ideal at a time when elites were consumed by fin-de-siecle anxieties about the fate of 'American' identity" ("Recognizing" 62). Finnegan again invokes the tools of rhetoric when acknowledging that, "in the nineteenth century, portraits were thought to be *ekphrastic*—that is, they were thought to reveal or bring before the eyes something vital and almost mysterious about their subjects" ("Recognizing" 68). By juxtaposing her analysis of the history of how the image itself came to be reproduced and published in *McClure's* with a historically and socially contextualized analysis of readers' written responses to the photo, Finnegan demonstrates how the complex interplay of image and text can open up new possibilities for the creation of rhetorical histories, one that "illustrat[es] how visual rhetoric constitutes a powerful world-making discourse" and a viable mode of inquiry ("Recognizing Lincoln" 74).

Hariman and Lucaites's extensive work with iconic photographs provides additional examples of visual rhetorical analysis that speak to Finnegan's criteria for the visual rhetoric project. Specifically, their analysis of the photo of the flag raising at Iwo Jima and its subsequent reproduction in society, as well as its echoes in the more recent image of the three firefighters raising the American flag after the events of September 11, 2001 not only provides a helpful example of important work in visual rhetoric but also allows us to transition into a discussion of the material and spatial components of visual rhetoric.

Like Finnegan, Hariman and Lucaites are concerned with the ways in which visual artifacts draw upon, communicate, and reproduce social knowledge. More specifically, they are concerned with the connections between iconic photographs and the shaping of public opinion and specific events, not only during the time of a particular photo's publication but over the course of its subsequent reception within society (11). They define the photojournalistic icon as "those photographic images appearing in print, electronic, or digital media that are widely recognized and remembered, are understood to be representations of historically significant events, activate strong emotional identification or response, and are reproduced across a range of media, genres, or topics" (Hariman and Lucaites 27). While they note on the one hand that "few images meet these criteria" (27) and thus provide detailed rationale for the set of images they choose to analyze in *No Caption Needed,* all of which they feel have had "distinctive influence on public opinion," they also acknowledge that "claims regarding influence are notoriously difficult to prove," and thus leave the door open for further identification and analysis of iconic imagery (7). Their study of the iconic photo of the Iwo Jima flag raising, which they see as "unquestionably the most popular image of World War II," exemplifies and fulfills the criteria for visual rhetoric projects (Hariman and Lucaites 21).

Hariman and Lucaites understand the Iwo Jima photo, which depicts six U.S. soldiers raising the American flag at the top of Mount Suribachi during the Battle of Iwo Jima, relative to the questions of rhetorical studies when they analyze the photo's composition, its appropriation over time, and its reproduction within popular media. Similar to Finnegan's work with the Lincoln photograph, they begin by contextualizing the photo's original publication on February 25, 1945 on the front page of newspapers nationwide (93). They then go on to describe how the photo's meaning has shifted over time and across generations (105). In the original photo, they eloquently analyze the significance of the men's poses and the meaning conveyed through the photo's visual composition of their coordinated efforts to plant the flag pole in the ground atop Mount Suribachi:

> There is a palpable harmony to the bodies as they strain together in the athleticism of physical work. Although the poses shift from being bent close to the ground to bearing a load to lifting upwards, one can draw a horizontal line across their belt lines, their knees all move together as if marching in step, all their physical energy flows along their common line of sight to the single point of impact in the earth. [. . .] We see the sure coordination of bodies with each other and with an instrument dedicated to their task. (Hariman and Lucaites 96)

Through their visual analysis of the image, Hariman and Lucaites are able to attribute its iconic power to "the combination of historical setting, visual transparency, and selfless action" (97). The complexity of the photo's power, they feel, is rooted in the "three codes of American public culture" conveyed through its composition (97). The anonymous but collective work ethic of the men working together conveys a sense of egalitarianism, while the flag raising conveys a sense of nationalism in its symbolism of "the nation's sacrifice and victory in World War II" (97). Finally, the "photo as a whole has the aesthetic quality of a sculpture" (97) thus affording it with a sense of civic republicanism; that is, the photo's monumental or sculpture-like quality affords it with the commemorative function often associated with public monuments and art installations, many of which represent civic virtue or political successes (101). In fact, as Hariman and Lucaites note, Congress later passed a bill subsidizing a memorial based on the photo; the memorial was unveiled in 1954 and resides outside Arlington National Cemetery (94).

The Iwo Jima photo's "compositional richness" (98) helps account for its subsequent appropriations within society over time. Like an original photograph or visual artifact, reproductions and appropriations of original works rely on social knowledge and reflect shifting contexts, and so the study of appropriations or reproductions that involve "copying, imitating, [or] satirizing" can make important contributions to visual rhetoric projects. Moreover, like the Iwo Jima Memorial at Arlington National Cemetery, appropriations need not take the form of solely textual media or visual photographs—they may involve sculpture, public art installations, or other forms of memorializing. While some reproductions retain the direct meaning of the original, as does the Iwo Jima Memorial, for example, others may stray from the original relationship between image and context, arguably creating a mystification of sorts. One of the more recent indirect appropriations of the Iwo Jima photo, for example, occurred just following the events of September 11, 2001, when reporter Thomas E. Franklin shot the now-iconic photo of the three firefighters raising the flag at ground zero.[4]

Aptly described by Hariman and Lucaites as a "profoundly visual event," the terrorist attacks on the World Trade Center in Manhattan and the Pentagon in Washington, D.C. were immediately followed by a huge, multimodal media response that served as a "direct and comprehensive emotional response to the event," joining "images of the destruction with depictions of the emotional reactions of ordinary people" (128). These large scale visual representations of responses to the disaster helped portray and constitute the public "as a unified nation whose civic virtue guaranteed triumph over disaster" (Hariman and Lucaites 128). One outcome of these visual narrative representations was that, by the close of the week, as Hariman and Lucaites put it, "a nationwide flag mania was underway" (128).

The public was soon inundated with images of the American flag both in the media and in everyday life; these images came to represent "fear and anger" on the one hand, and instances of patriotism and "civic pride" on the other (128). Soon, a photographic icon emerged from these images: that of the three firefighters raising the U.S. flag at the site of what was just formerly the World Trade Center. The public immediately recognized the image as reminiscent of the Iwo Jima photo. Even Thomas Franklin, the photographer who took the photo, saw the firefighters raising the flag "and thought, 'Iwo Jima'" (Hill and Helmers 5). As Charles Hill and Marguerite Helmers describe, this photo demonstrates an instance of intertextuality, "the recognition and referencing of images from one scene to another" (5). Appropriations, for example, rely on intertextuality for their reception and tap into social knowledge in their ability to construct connections from one context to another. While the photo of the firefighters is now officially referred to as "Ground Zero Spirit," a version of the image was initially captioned by *People* magazine as "'an echo of Iwo Jima'" (Hariman and Lucaites 131). Like the Iwo Jima photo, say Hariman and Lucaites, the firefighters in the photo are "dominated by their anonymity and working class norms of hard physical labor, self-sacrifice, and loyalty" (132). In addition, like the Iwo Jima photo, the flag pole "cuts across the frame on the same diagonal [. . .] while the flag itself is moved upward by coordinated effort" (133). The background is bleak and empty and, like the Iwo Jima photo, the image itself does not depict war in progress, though "precipitating events and surrounding discourse might suggest otherwise" (133). While the context has shifted, the result is again the reflection of the American codes of public culture: "egalitarianism, nationalism, and civic republicanism" (Hariman and Lucaites 133). On the one hand, while a pervasive symbol such as the American flag carries with it particular "ideological formations such as nationalism," it is also open to appropriation and serves not only as a means for perpetuating a particular normative mode but also allows for "inflection and critique" (135). Finnegan describes the way in which viewers bring their own contexts to bear on the interpretation of images. Likewise, one goal of Hariman and Lucaites's discussion of the Ground Zero photo is to show how iconic photographs, rather than adhering to a "fixed meaning," function as malleable resources that invite the public to construct connections and "coordinate available structures of identification within a performative space open to continued and varied articulation" (135).

Moreover, these articulations and appropriations need not be purely or traditionally visual in nature, and need not take the form of solely print media or photographs. That is, just as the Iwo Jima Memorial emerged as a direct appropriation of the original photo, following the emergence of the iconic Ground Zero Spirit photo, "there was an immediate call to establish a memorial park at ground zero in Manhattan that would include a statue of the firefighters raising the flag" (Hariman and Lucaites 129). Certainly, the events of September 11 set into effect their own course of memorializing.

In the days, weeks, and months following the attacks, the public witnessed and participated in various acts of memorializing that spanned modes of visual and material representation. Such activities included the creation of makeshift memorial spaces constructed by citizens, media documentaries and reports, and other objects of print and popular media. As L.J. Nicoletti describes, we were "consuming political rhetoric and visual forms of memorialization as never before" (56). To help her first-year writing students cope with the events of September 11 and respond to the types of memorialization they were witnessing in the mass media, for example, she developed an assignment in which her students designed their own memorials, thus enabling them to take part in what she calls the "language of memorial spaces" (56). While this is not a book about the rhetorics of public memory per se, it is clear that an interest in memorializing or commemoration often serves as a catalyst for the creation of many public sculptures, exhibits, and other multimodal displays or artifacts. When planning their own memorials in response to the events of September 11, for example, Nicoletti asked her students to consider design components such as "symbolism, setting, audience, scale, permanence, and inscription," thus reflecting the idea that appropriation, intertextuality, and social knowledge are important components of commemoration (56). Moreover, the consideration of design elements such as scale and permanence speaks to the physicality and spatiality of rhetorical artifacts that are not only visual but also material in composition. Visual artifacts that are also tangible and spatial, and invite engagement not only with the mind but with the whole body, can then be understood as objects of material rhetoric.

Like projects of visual rhetoric, material rhetoric seeks to understand physical artifacts and sites such as memorials, parks, or green spaces, and even multimodal artifacts such as the GPS in the context of the questions of rhetorical theory, while also reconsidering rhetorical theory relative to the challenges brought about by the study of materially rhetorical artifacts. Like the projects of visual rhetoric, material rhetoric too seeks to uncover the power and knowledge dynamics related to the study of rhetorical artifacts that incorporate visual, textual, physical, spatial, or other multimodal components. The relationship between these different generic modes is likewise seen as inviting new interpretive possibilities. Finally, building on those criteria of visual rhetoric projects, and fully compatible with them, material rhetoric seeks to more explicitly understand the influence of rhetoric on the body.

As chapter two describes in more depth, material rhetorical analysis has at its core a focus on the impact of rhetorical artifacts on contextualized, bodily experience. Carole Blair's approach to understanding material rhetoric, one that underpins much of the work of this book, provides a toolkit for analyzing visual texts that also have material and spatial components. In her landmark essay and important project of material rhetoric, "U.S. Memorial Sites as Exemplars of Rhetoric's Materiality," she begins from the point of acknowledging that within the field of rhetoric, "we lack an idiom for referencing talk, writing, or even inscribed stone as material"—that we struggle with "the lack of a materialist language about discourse" (17). To better understand rhetoric's materiality, she examines five memorial sites: the Vietnam Veteran's Memorial, the AIDS Memorial Quilt, the Civil Rights Memorial, Kent State University's May 4 Memorial, and the Witch Trials Tercentenary Memorial.

Within the context of her study of these five memorials, which she sees as "not necessarily representative of all memorials" but rather as able to reveal unique ideas about material rhetoric, Blair poses five questions that help to redefine what counts as a text (24). She asks, for example: "(1) What is the significance of the text's material existence? (2) What are the apparatuses and degrees of durability of displayed by the text? (3) What are the text's modes or possibilities of reproduction or preservation? (4) What does the text do to (or with, or against) other texts? (5) How does the text act on people?" (30). Blair describes the ways in which a text's physical composition will affect its durability, vulnerability, and possibilities for modes of preservation and reproduction. She compares, for example, the black granite of the Vietnam Veterans Memorial (VVM) with the relative vulnerability of the fabric of the AIDS Memorial Quilt, and the types of spaces engendered by the types of interactions that each invites. She also describes also how the use of black granite has become appropriated and has subsequently been reproduced indirectly in other memorials. For instance, to best understand the Kent State Memorial (KSM), it is first necessary to know that the VVM is composed of black granite. At the KSM, visitors follow a prescribed path of black granite into the memorial. The black granite composition of the KSM implicitly describes the shootings as embedded within the context of the Vietnam War. The physical composition of the VVM, then, not only becomes prerequisite knowledge for understanding the KSM but also encourages an intertextual reading of it. Similar to the intertextual relationships between visual texts, such as that of the Iwo Jima and Ground Zero photos, one consequence of the appropriation of physical features of material texts is that they begin to develop their own textual and intertextual identities over time. Blair describes how our reading and bodily experience of material texts can shape our perceptions of the events they represent in ways that allow for a fuller understanding of the consequences of these events on the mind and body. By exploring the material aspects of a text's durability, modes of reproduction, and visibility, Blair not only implicitly builds on the goals of visual rhetoric projects but also taps into the materialist language about discourse that has for so long been a missing component of rhetorical analysis.

The study of visual rhetoric, then, when also understood in light of the questions posed by material rhetorical analysis, has the potential to help illuminate the spatial components of texts, places, and other physical artifacts. The idea of a rhetorical approach that merges visuality and spatiality is greatly appealing, for as someone particularly attentive to the intersections of rhetoric and geography, I have often understood visual and material artifacts largely through both a rhetorical and a geographical lens—as discursive objects that facilitate spatial understanding, are situated in time and space, and make important claims to knowledge. [5] Again, visual and material artifacts may include physical sites such as factories, public monuments, or art installations that function commemoratively to reflect or perform particular cultural moments, often guiding both the mind and body toward specific interpretations. They can also include multimodal technological devices such as global positioning systems (GPS), which, through their physicality, use of audio and visual cues, and the cartographic texts they produce, can function rhetorically to make specific knowledge claims, influence bodily practices, or guide movements and decision-making. To understand visual rhetoric as also concerned with studies of space, the body, and materiality will, as I argue in this book, allow us to more fully understand the broader implications and consequences of the rhetorical work of visual artifacts in the world. In short, it will allow for a more inclusive understanding of the sort of work that projects of visual rhetoric can accomplish.

In this book, I aim to cast a brighter light on the important connections between visual rhetoric, material rhetoric, space, and bodies, in order to show the value of these connections within and beyond the field of rhetoric; ultimately, I aim to create a unique space for material rhetorics along the spectrum of what I envision to be a visual-material rhetoric. Again, when I refer here to *rhetoric,* I am describing the idea that texts, artifacts, and discourses are "partisan, meaningful, and influential," to the extent that they have the capacity for consequence and may influence our understandings of specific contexts in ways that impact both the mind and body (Blair 72, "Reading"). Compatible with this notion, *material rhetoric* considers the text from the perspective of its relative durability and reproducibility, its ability to work with and against other texts, and most important, its ability to understand space and place as rhetorical and as affecting both the mind and body.

When I use the terms *space* and *place* in this book, I borrow from Michel de Certeau's ideas about how a particular space is composed of mobile elements that are in constant motion and relationship with each other. Subsequently, the interactions of texts, artifacts, bodies, and discourses within that space constitute a more specific sense of *place* as they move and interact with and against each other within particular contexts and configurations. Given this understanding, *place* may be viewed as happening within a *space*. Place, writes de Certeau, "is thus an instantaneous configuration of positions" (117). For example, the parks and public commemorative sculptures at the Lowell Mills National Historical Park in Lowell, Massachusetts (the object of focus in chapter three of this book), may foster a specific sense of place largely because of the interpretation invited by their layout and constituted by the activities and movements of park visitors. Likewise, specific representations of the park as performed through the park map, or the features deployed within the map, may also be seen as constructing a more nuanced version of a place. To view a space as rhetorical, then, is to acknowledge the capacity for consequence borne out of the interaction of the texts, artifacts, bodies, and discourses deployed within it, and the sense of place engendered by those interactions. With these ideas in mind, this book progresses along a contextual continuum that explores and tests the value of visual-material rhetorics from three interconnected perspectives: that of the human body, the posthuman body, and the nonhuman body.

In chapter one, I begin by telling the story of visual rhetoric largely from the perspectives of geography and space. Specifically, I see visual rhetoric through the lens of *critical cartography,* a subdiscipline of cartography that sees geographic knowledge as tied to power relations, and understands mapping and the practices of visuality as informed by cultural contexts (Crampton and Krygier 11). Here, I situate visual rhetoric as able to help account for the spatial dimensions of texts, and I subsequently describe my understanding of space as rhetorical. I describe more specifically how maps function as rhetorical artifacts. Then, in order to better contextualize the intersections of visual rhetoric and critical cartography, I provide a brief analysis of *Photo AS17–148–22727* (commonly referred to as photo 22727), the famous NASA photo of the whole earth taken by the crew of Apollo 17 in December 1972. As a visual artifact that functions as both iconic photograph and cartographic representation, an analysis of photo 22727 helps pave the road for a discussion of how visual rhetoric can be more attentive to the relationships between materiality, space, and the body. I then move on to identify some possible limitations in an understanding of visual rhetoric that does not explicitly consider a text's influence on the body, and thus call for a mode of interaction with the text that more explicitly does so. Within the context of this call to action, I introduce Blair's theory of material rhetoric in more detail and Michel Foucault's theory of heterotopias, and I understand their ideas as providing a point of entry into a more embodied approach to visual-material rhetorics. As Katherine Hayles describes the concept, embodiment understands bodily experience as imbricated in and shaped by specific social, cultural, and temporal contexts (*Posthuman*). Because contextual experiences shape and perpetuate ways of knowing, embodiment may also be understood as contributing to embodied knowledge, or ways of knowing and discursive practices that are informed, perpetuated, and sustained by contextualized, bodily experience.

In chapter two, I take a closer look at discussions related more specifically to material rhetoric, situating among them Blair's (1999) theory of material rhetoric and Foucault's (1986) concept of heterotopias. Briefly put, a heterotopia may be understood as a heterogeneous, contested space that nonetheless includes identifying characteristics specific to that space. For example, the maps, green spaces, and sculptures at the Lowell Mills National Historical Park work with and against each other to represent and perform specific moments in Lowell's history. Combining and extending the theoretical approaches of Blair and Foucault not only allows for an emphasis on spatiality as the common thread linking visual and material rhetorics, but also provides the foundation for what I then call a visual-material rhetorical approach, one that not only accounts for the multimodal, spatially-situated artifact but is also mindful of its impact on the embodied subject. With these ideas in mind, the following three chapters work to better understand how a visual-material rhetorical approach that accounts for embodiment can illuminate the rhetorical situation, or more pointedly, how conceiving of visual-material artifacts as rhetorical and embodied reveals a more nuanced understanding of the particular moments, events, and debates—and the consequences of each—that these artifacts set out to represent.

In chapter three, I account for the impact of visual-material rhetorics on contextualized, bodily experience. I draw from a combination of observation, interviewing, and archival research to conduct a visual-material rhetorical analysis of the maps, wayfinding devices, green spaces, and public commemorative sculptures at the Lowell Mills Park. The chapter not only demonstrates how the park's spatially-focused, cultural artifacts engage visitors and facilitate their navigational decision-making, but also how these artifacts reflect and perform the impacts of the mills on the lives of the Mill Girls who labored there in the early 1800s. I demonstrate how a visual-material rhetorical approach can help create room for more nuanced and empathetic understanding of the contexts in which these women lived and worked, thereby allowing audiences to engage with greater empathy in the lives and struggles of tangential groups, such as, in this case, the lives of the Mill Girls.

In chapter four, I again explore the ways in which visual-material rhetorics can help account for contextualized, embodied experience, this time considering the rhetorical actions of drivers who navigate their routes with the assistance of GPS devices. Drawing on interviews with GPS users, I explore the ways in which GPS technology helps mediate particular experiences, or how GPS users work with and against the technology to make purposeful decisions that sometimes foster, and sometimes constrain spatial understanding. In this sense, I see visual-material rhetorics as able to function in the service of advocacy—to function rhetorically in such a way that enhances informed decision-making and influences our capacity to understand our worlds. A visual-material rhetorical approach helps reveal the processes that shape these levels of interaction and their implications for the technologically-mediated, posthuman body. Hayles writes that we ought to understand the posthuman body as able to interact with information technologies without necessarily being seduced into "fantasies of unlimited power and disembodied immortality"—that we must understand modes of being as inextricably linked to the material world (*Posthuman* 5). Likewise, for Collin Brooke, a posthuman rhetoric entails "a return to embodied information, [and] involves a revaluing of partiality" (791). Moreover, I would add that understandings of the posthuman body and the idea of a posthuman rhetoric need not be constrained solely to the consequences of rhetorical action on human animals. For, as Cary Wolfe notes, our understanding of the posthuman is also tied to larger issues of "nonhuman modes of being" (193). Given these views, then, understandings of the posthuman body may extend beyond human ways of knowing to the consequences of rhetorical action on nonhuman animals such as, for example, marine mammals, as chapter five helps show.

In chapter five, I show how a visual-material rhetorical approach can help advocate for the bodies of nonhuman animals who may otherwise be unable to advocate for themselves within a specific rhetorical context. Set within the context of a recent federal court case between the Natural Resources Defense Council (NRDC) and the National Marine Fisheries Service (NMFS), this chapter explores how the map, as an artifact of visual-material rhetoric, can function as a mediating device in policy-making, or as a persuasive tool in debates between institutions. In this case, the map functions rhetorically to illuminate competing knowledge claims about contested space in order to influence a contemporary debate about environmental policy and marine mammal protection, and advocate for nonhuman bodies who cannot be physically present in the setting of the debate. From a rhetorical perspective, then, I understand advocacy both as referring to the ways in which texts, artifacts, and discourses have the potential to promote new knowledge among certain groups or individuals, as well as support or promote the interests of tangential voices, underrepresented groups, or groups who may otherwise be unable to advocate for themselves within the setting of a particular debate.

Finally, chapter six concludes the book by considering how these three cases, when understood together, demonstrate the implications of visual-material rhetorics for the body, the posthuman body, and the nonhuman body. Viewed holistically, I show how these cases and the theories underpinning their analysis help demonstrate how visual-material rhetorics can illuminate the contexts that shape our various lived and embodied experiences. I discuss the implications of visual-material rhetorics for rhetorical analysis and understanding rhetoric as advocacy work, as well as consider future directions for research and undergraduate and graduate pedagogy. Subsequently, as I hope readers will find, visual-material rhetorics, when understood as embodied knowledge, can work across rhetorical situations in the service of advocacy to constitute a sustainable project of inquiry.

LOCATING VISUAL-MATERIAL RHETORICS

1 Visual Rhetoric and Spatiality

In this chapter, I tell the story of visual rhetoric largely from the perspective of critical cartography, a subdiscipline of cartography that sees geographic knowledge as tied to power relations and understands mapping and the practices of visuality as informed by cultural contexts (Crampton and Krygier 11). A story is of course a subjective account, and I have been interested here in crafting a narrative that demonstrates how visual rhetorics, when considered through a geographical lens, can help account for the spatial dimensions of texts and artifacts, and subsequently allow for an understanding of space as rhetorical, and of rhetoric as spatial. After describing more specifically how maps function as rhetorical artifacts, I contextualize the intersections of visual rhetoric and critical cartography by providing a brief analysis of photo 22727, the famous NASA photo of the whole earth taken by the crew of Apollo 17 in December 1972. As a visual artifact that functions as both iconic photograph and cartographic representation, an analysis of photo 22727 not only helps situate visual rhetoric relative to critical cartography but also paves the way for a discussion of how visual rhetoric can be more attentive to the relationships between materiality, space, and the body. This discussion of visual rhetoric and spatiality, however, is a means to a greater end, for I link the two in order to suggest that there exist some possible limitations in an understanding of visual rhetoric that does not explicitly consider an artifact's contextualized engagement with the body. I thus call for a visual rhetoric that more expressly accounts not only for a rhetorical artifact's material and spatial components but also for its subsequent impact on the body. Within the context of this call to action, I introduce Blair's theory of material rhetoric and Foucault's theory of heterotopias, which, in chapter two, I then describe in more detail as helping constitute a sustainable theory of visual-material rhetorics that provides a point of entry into a more embodied rhetorical approach—

one that can help demonstrate the value of visual-material rhetorics both within and beyond the field of rhetoric.

VISUAL RHETORIC AS A PROJECT OF INQUIRY

As described in the introduction, my interest in the intersections of rhetoric, visual studies, geography, and critical cartography has allowed me to arrive at an understanding of visual rhetoric compatible with what Cara A. Finnegan describes as "a *project of inquiry,* rather than a product" ("Review Essay" 244). Again, this conception allows for two related research trajectories within visual rhetoric (244). These trajectories function 1) to focus on the study of the artifact itself, and 2) to explore the significance of visuality for rhetorical theory. Finnegan feels we need to be more mindful of how the practices of visuality influence, affect, or shift understandings of rhetoric (245). Visual rhetorical analysis, she writes, "should recognize the influence of visual artifacts and practices, but also place them in the contexts of their circulation in a discursive field conceived neither as exclusively textual nor exclusively visual" (245).

The question of how to work analytically with visual artifacts should not be minimized, though at the same time it should not be cause for paralysis either. While scholarship in visual rhetoric has rightly begun to acknowledge the interplay of the verbal and the visual, moving past older references to the verbal/visual dichotomy[1], acknowledgement of the interplay between textual, visual, and material ways of knowing does not make for a methodological non-issue. In fact, to acknowledge this interplay and its attendant multimodal epistemologies invites a new set of methodological questions. On the one hand, as Finnegan notes, "labeling as 'visual rhetoric' artifacts such as photographs, memorials, art, images, and advertisements creates a false category," in that many of these artifacts include textual or linguistic characteristics (244). Thus, to place such artifacts under the rubric of visual rhetoric "ignores the often untenable distinction between the visual and verbal in practical discourse" (244). On the other hand, visual and material artifacts arguably require specific methodological treatment, thus necessitating that we acknowledge such categorical and analytical distinctions within our analyses.[2] I suggest that it is possible to understand the textual, visual, or material qualities of rhetorical artifacts as functioning along a spectrum, without necessarily creating the false categories about which Finnegan cautions, so long as we are attendant to the nuanced readings that such artifacts require. Such mindfulness will only serve to further our understandings of how multimodal contexts and artifacts can influence, affect, or shift our understandings of rhetoric.

Moreover, the idea that the study of visual artifacts and practices should be contextualized within their discursive field is, to my mind, not incompatible with an understanding of visual rhetoric as always already concerned with embodied practice. In *How We Became Posthuman,* for example, N. Katherine Hayles describes the idea of embodiment as related to but different from the body; the difference, she feels, is linked to a consideration of the cultural contexts in which the subject is situated:

> Embodiment differs from the concept of the body in that the body is always normative relative to some set of criteria [. . .] In contrast to the body, embodiment is contextual, enmeshed within the specifics of place, time, physiology, and culture, which together compose enactment. [. . .] Experiences of embodiment, far from existing apart from culture, are always imbricated within it. (196–97)

To understand visual rhetorics and the rhetorical study of visual, spatial artifacts as embodied practice not only allows us to more explicitly consider the contextualizing features of "place, time, physiology, and culture" that Hayles describes but also requires that we understand the visual as concerned with space, place, and the body (196). In other words, to fully understand the broader implications and consequences of the rhetorical work of visual and material artifacts in the world, we must understand visual rhetorics as also concerned with and receptive to studies of space, place, and the body.

SPACE AS RHETORICAL

To engage in a study of visual-material rhetorics from a vantage point that understands space as rhetorical requires a brief review of how space and place may be understood to function in this book. As mentioned earlier, I borrow from the ideas of philosopher Michel de Certeau, who has familiarized the notion that "space is a practiced place" (117). *Space,* he notes, "is composed of intersections of mobile elements. It is in a sense actuated by the ensemble of movements deployed within it. [. . .] Thus the street geometrically defined by urban planning is transformed into a space by its walkers" (117). These "walkers," or the subjects residing within a space, also give that space its sense of "place." Subsequently, *place* is concerned with relationships among the elements within the space and the ways in which they interact and coexist (117). In this sense, space and place function together and may be understood as constituting the power relations that make possible particular ensembles of movements and intersections of mobile elements. Again, de Certeau understands a place as "an instantaneous configuration of positions" (117). For example, the green spaces and public commemorative sculptures at sites such as the Lowell Mills Park (the subject of chapter three) may foster a specific sense of place largely because of the interactivity invited by their layout and subsequently by the activity of their "walkers," or the movements of visitors within the park. Likewise, specific representations of the park as performed through the park map, or the features deployed within the map, may be seen as constructing more nuanced versions of a place. Human geographers have also suggested that certain places engender a specific sense of place, or particular feelings or emotions associated with a place (McDowell and Sharp 210).[3] To view a space as rhetorical is to acknowledge the capacity for consequence borne out of the interaction of the texts, artifacts, bodies, and discourses deployed within it, and the sense of place engendered by those interactions.

THE MAP CAN TAKE US FROM HERE TO THERE

The idea that space and place are socially produced and contextually relevant has implications for cartography as well. A discussion of cartographic representation serves as an ideal point of entry into understanding how visual rhetoric can be more attentive to the relationships between space and the body. Scholars aligned with critical cartography, for example, not only take as given that the cultural work of the map relies on multimodality and intertexuality but they also understand the map as rhetorical, and as always already shaping and shaped by the cultural contexts in which it is immersed. Contemporary cartographic practice has largely begun to acknowledge that mapping, while historically understood as an objective, scientific practice,[4] is also a cultural practice that may impact "how the space is perceived and what action takes place within it"; in other words, mapping may also "represent an exercise in power" (McDowell and Sharp 25). As geographers Jeremy Crampton and John Krygier note, mapping invites participation and dialogue: "If the map is a specific set of power-knowledge claims, then not only the state but others could make competing and equally powerful claims" (12). Moreover, they write, "critical cartography assumes that maps *make* reality as much as they represent it" (15). Critical cartographers then understand mapping as an active practice that can shape knowledge, reflect power dynamics, and serve as a means for advancing social change (15). One type of mapping made possible through a critique of critical cartography is what Crampton and Krygier refer to as "everyday mappings" (25). Everyday mappings may be "experiential or narrative, and creatively illuminate the role of space in people's lives by countering generalized and global perspectives" (25). The digital maps created by GPS devices, for example, fit the bill well; as chapter four will discuss in more detail, they are multimodal, rhetorical, everyday texts created in the moment by users who want tailored information about their immediate environments. These cartographic texts, much like the work of other spatial artifacts and representations, have both an immediate impact on contextualized, bodily experience as well as broader consequences within and beyond the rhetorical situation. Other mappings may resemble more traditional modes of visual representation such as photographs, more so than what we might typically consider to be a map. In the discussion that follows, I first describe more specifically how maps function as rhetorical artifacts through their potential for visual and textual interplay, selectivity, and modes of projection, as well as the ways in which they are always already implicated in cultural practice. Next, to better contextualize the intersections of visual rhetoric and critical cartography, I provide a brief analysis of a visual artifact that counts as both iconic photograph and map: Photo 22727, also known as the "Blue Marble." An analysis of this image not only helps show the connections between visual rhetoric and critical cartography but also begins to demonstrate how visual rhetoric can be more attentive to the relationships between materiality, space, and the body.

The Map as Rhetorical Artifact

The late geographer J.B. Harley conveys an understanding of the map as rhetorical and able to present arguments about the world when he writes: "My position is to accept that rhetoric is part of the way all texts work and that all maps are rhetorical texts. [. . .] All maps strive to frame their message in the context of an audience. All maps state an argument about the world [. . .]. All maps employ the common devices of rhetoric such as invocations of authority" (242). While Harley understands the map as contextually-specific and as requiring a specific audience and purpose, his work is sometimes critiqued for its more inward focus on the production of the map itself, rather than on the "nuanced and multiform" processes of mapping and the social and political contexts that inform "the production of geographical images" (Pickles, *A History of Spaces* 146). In describing cartography's recent turn to "processual" modes of knowing, Leila Harris and Helen Hazen advocate for a focus on the "multiple, reiterative production and reproduction of maps as they are engaged in multiple times and spaces," rather than focusing solely on the power dynamics that inform the production of specific maps (51).

Interplay of Text and Image.

In rethinking the ways in which mapping is informed by specific contexts and relationships, Harris and Hazen also note that "key insights are possible by analyzing the ways that lines and colours *become* maps, are given meaning, and are performed in relation to specific knowledges or techniques, or through relational engagements involving mapmakers or users" (51). The idea that the graphical features of the map not only shape its meaning but are also informed by the cultural contexts and relational processes in which mapmakers and users are immersed broaches an understanding of the map as both sign system and cultural artifact.

Noted cartographers Denis Wood and John Fels likewise understand the cultural work of the map but also acknowledge its fundamental composition as a sign system that is comprised of both word and image when they describe a map's meaning as tied to the interplay of visual and textual elements inherent in its display: "As word lends icon access to the semantic field of its culture, icon invites word to realize its expressive potentials in the visual field. The result is the dual signification virtually synonymous with maps, and the complementary exchange of meaning that it engenders" (Wood and Fels, "Designs on Signs" 80). Acknowledging the semiotic components of the map, cartographer David Turnbull notes that cartographic representation generally falls under two main categories: *iconic* and *symbolic*. Aligned with Peirceian semiotics, Turnbull understands iconic representation as bearing a direct likeness to the feature it describes; it attempts "to directly portray certain visual aspects of the piece of territory in question," whereas a symbolic representation taps into social contexts in order to make meaning, and makes use of "purely conventional signs and symbols, like letters, numbers, or graphic devices" (3). Many Western maps employ both iconic and symbolic features; however, this is not to say they explicitly distinguish between the two modes of representation. Rather, these two modes are common and implicit functions of cartographic convention and representation.[5]

Selectivity

Similarly, Ben Barton and Marthalee Barton note that part of how the map creates meaning is through its selectivity, or through the inclusion and exclusion of information (55). Lawrence Prelli also demonstrates the notion of the rhetorical selectivity of display in his analysis of the Georges Bank and the boundary line dividing "United States and Canadian jurisdiction over resources in the Gulf of Maine" (90). At stake in the debate over these boundary lines was control of the lucrative fisheries in the region (90). Here, Prelli examines how maps and graphics were used selectively by both parties to influence "how the gulf's features were seen and disposed the attitudes of those who saw them" (91). In doing so, Prelli explores the idea of "visual *taxis,*" or how visual artifacts may be implemented in the strategic structuring of an argument "for maximum persuasive effect with particular, targeted audiences" (92).[6] As Barton and Barton and Prelli emphasize, selectivity is clearly a large component of visual and material representation. Turnbull too agrees that while the map cannot possibly account for or "display all there is to know about any given piece of the environment," for a visual representation of space to be deemed a map, it "must directly represent at least *some* aspects of the landscape" (3).

Scale and Projection

In addition to understanding selectivity as a component of cartography's epistemic capacity, projection and scale also shape how the map conveys particular meaning. Maps rely on scale to "bring the worldview to manageable proportions" (Dorling and Fairbairn 25). Maps that represent the whole earth on a single piece of paper or on a computer screen, for example, are "small-scale" maps, because they convey relatively little detail about a vast area within a small space. By contrast, a map of a city park that portrays "the landscape, other spatial features and their variation in great detail over a limited tract of space" can be considered a "large-scale" map, because a unit of measurement such as 1 centimeter on the map may be equivalent to 5 meters on the ground (25). What counts as large-scale or small-scale more precisely, however, is "subject to enormous subjective individual variation," and so these terms are not generally understood as conveying precision (25).

The idea of projection helps account for how "the irregularity of the earth's surface can be precisely addressed on a two-dimensional plane" (Dorling and Fairbairn 25). Barton and Barton describe projection schemes like the Mercator view as potentially sustaining visual distortions that "are embodied in the cartographic space as a grid" (58). Thus, they view the grid as an ideologically-charged representational device that has a propensity toward distortion, despite the fact that its purported goal is to convey accurate models of the terrain by positioning space along *equal* lines of latitude and longitude (58). Turnbull also points out that the grid is socially constructed and that it does not correspond with a specific physical reality or territory (26). Thus, a generic convention of the map in Western culture is its imposition of the grid onto the landscape it represents. Like Barton and Barton, Turnbull and geographer Mark Monmonier both note the potential distortions that may result from the use of various map projections. The round Earth cannot be projected onto a flat, two-dimensional surface without some level of distortion; as a result, Turnbull says, various projections have been devised to account for this issue (6). While "no one projection is the best or the most accurate," different types of projections have different purposes for which they are more or less well-suited: "A particular projection is selected by the mapmaker on the basis of functional and perhaps aesthetic criteria, or because of a specification or convention" (Turnbull 6). Monmonier expresses a wariness when describing distortions resulting from the Mercator projection, which he feels is a "demonstrably bad choice" in projection for any map "not related to navigation" (*Mapping it Out* 53). As chapter five describes in greater context, the Mercator projection is most useful in sea navigation, wherein a straight line represents the actual compass bearing. This projection, however, "so grossly distorts areas and distances that the poles lie off the map at infinity" (Monmonier, *Mapping it Out* 48). No other projection, Monmonier feels, has been "so abused in the pursuit of size distortion" (*How to Lie with Maps* 94).

Maps Constitute Ways of Seeing

Maps are thus context-bound and create meaning through their selectivity, their use of particular cartographic conventions, their imposition of the grid, the expectation that at least *some* aspects of the landscape are represented, and their use of both iconic and symbolic features. How, then, may the map be defined? For Turnbull, "[m]aps are graphic representations that facilitate a spatial understanding of things, concepts, conditions, processes, or events in the human world. [. . .] [The map is a] graphic representation of the milieu, containing both pictorial (or iconic) and non-pictorial elements" (3). Maps are then partial, selective representations of the world; they are always in flux and respond to their shifting contexts and relations. Their use of lines, colors, and other graphical features are likewise responses to particular social and cultural contexts and "relational engagements" (Harris and Hazen 51). Crampton also views the act of mapping as a relational cultural practice, one that needs to look outside of itself as much as it looks within; in other words, he is interested in the contexts and conditions that allow for different types of cartographic meaning to come into being (52).[7] Crampton understands the map not only from the vantage point of its work as an inherently ideological document but also as one that goes on to invite interpretation and various contextualized readings. These contextualized readings may happen outside of the discipline that produced the artifact, and while those disciplinary practices must not go unchecked, the cultural work of the map extends far beyond the site of its production to influence the material worlds and bodies that it represents. Pickles also understands cartographic practice as functioning beyond the production of the artifact itself (an idea supported by Finnegan as well), and as tied to the larger project of understanding how space influences embodied experience. Here, Pickles quotes from Denis Wood when he notes that "the practice of map use is not to send a message, but to bring about a change in the way another person, or group of people, see the world. It is 'out of their interaction in the social worlds they inhabit that people bring forth cultural products like maps'" (qtd. in Pickles, *A History of Spaces* 66). One example of a map that has arguably brought about a change in how people see the world is photo 22727.

THE RHETORICAL WORK OF NASA PHOTO
AS17–148–22727: THE BLUE MARBLE

NASA's photograph *AS17–148–22727*, taken during Apollo's final journey in December of 1972, is an ideal image through which to briefly demonstrate the connections between visual rhetoric and cartographic representation (Figure 1).

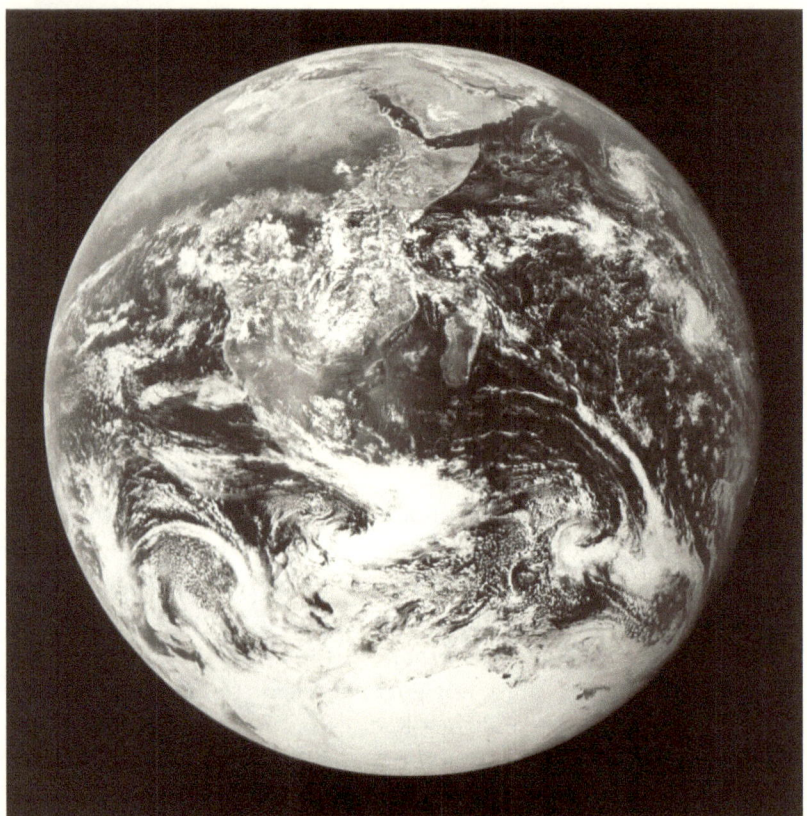

Figure 1: NASA Photo *AS17–148–22727*, 1972. "View of the Earth seen by the Apollo 17 crew traveling toward the moon." (Courtesy NASA and NSSDC Photo Gallery.)

The photo was taken by the astronauts aboard Apollo's final flight on December 7, 1972.[8] It was released by NASA on December 23, 1972 and was published on the front page of newspapers across the country over the Christmas weekend (Hartwell). The photo decenters Europe and privileges the Southern Hemisphere, thus working against the historically ethnocentric view of the globe that Monmonier has often critiqued (*How to Lie*). As geographer Denis Cosgrove has described in his study of photo 22727, the image depicts a "perfectly circular earth within a square frame [. . .]. The edges of the floating globe seem to dissolve into the surrounding black, an impression produced by the earth's atmosphere" (*Apollo's Eye* 260). The photo is predominantly composed of brown, white, and blue tones, which serve to "clearly define the landmasses of Africa and the Arabian peninsula, the South Atlantic and Indian Oceans, and the island continent of Antarctica" (260). The image functions as a complex artifact of visual rhetoric. Recognized by many as a widely reproduced, iconic photo, it is also a map of the whole earth and thus an object of cartographic practice. Further, the social and political contexts in which the image was situated when it was first introduced to the public in 1972 can help us understand the power and knowledge dynamics at work in its circulation and thus its rhetorical power as both iconic image and cartographic representation.

The Image as Both Iconic Photo and Cartographic Representation

Photo 22727 fits within the baseline criteria articulated by Hariman and Lucaites for what counts as an iconic image. Reproduced throughout the years in various forms of print and electronic media, the photo of the whole earth is "widely recognized and remembered," associated with the "historically significant" final flight of Apollo 17, and may be read as "activat[ing] strong emotional identification or response" in its audience (Hariman and Lucaites 27). Introduced during a period of time in which the environmental movement of the United States was just beginning to emerge, the image has become associated with the idea of environmentalism, has been appropriated by environmental groups, and has arguably shaped perceptions of how the earth is imagined within public discourse. Cosgrove helps us understand how the photo works rhetorically both as an iconic image and as a cartographic representation. On the one hand, he says, photo 22727 may be understood primarily as an iconic photo, for "the frequency with which photo 22727 is reproduced in reverse or inverted suggests that its status is iconic rather than cartographic. While it is instantly recognized as an image of the earth, few register its precise geographical contents. Most respond primarily to its cosmographic and elemental qualities" (*Apollo's Eye* 261). On the other hand, Cosgrove seems to implicitly understand the image as both iconic photo and object of cartographic practice when he notes that "the image's *geographical, compositional,* and *tonal* qualities give it unusually strong imaginative appeal, aesthetic balance, and formal harmony" (260, emphasis added). Moreover, if we consider Cosgrove's definition of the practice of mapping, we can see that an iconic, geographic image such as photo 22727 indeed accomplishes cultural work that influences our understanding of the world and shapes the geographic imagination: "to map is in one way or another to take the measure of a world [. . .] in such a way that it may be communicated between people, places or times. The measure of mapping is not restricted to the mathematical; it may equally be spiritual, political or moral" (Cosgrove, "Mapping Meaning" 2). Thus, in moving away from traditional notions of cartography as positing neutral, correct, relational models of the terrain, we can begin to understand how images of the earth like photo 22727 function as mappings that cultivate critical thought or reflection among its viewers. Likewise, Cosgrove understands mapping as a knowledge-making practice that encourages us to step outside of our traditionally held assumptions in order arrive at new imaginings of our world: "Acts of mapping are creative, sometimes anxious, moments of coming to knowledge of the world, and the map is both the spatial embodiment of knowledge and a stimulus to further cognitive engagements" (Cosgrove, "Mapping Meaning" 1). Indeed, interpretations of photo 22727 enable cognitive engagement with the idea of how we understand our world.

Ways of Seeing Photo 22727

Photo 22727 dramatically displays Earth as a singular entity, surreal and lacking the context of broader surroundings; it presents the viewer with the "whole, unshadowed globe floating in the blackness of space and given NASA number AS17–22727" (Cosgrove 257). From the perspective of the photo as a cartographic, rhetorical artifact, we might consider the power of its small scale, which seemingly marks a territory encompassing "the whole of creation": "In scale, mapping may trace a line or delimit and limn a territory of any length or size, from the whole of creation to its tiniest fragments; notions of shape and area are themselves in some respects a product of mapping processes" (Cosgrove, "Mapping Meaning" 2). The apparent vastness of the map's territory, its small scale, and its perfect circularity contribute to its feeling of disembodiment, and may initially spark interpretations along the register of what Cosgrove terms the "one-world discourse" (*Apollo's Eye* 263). That is, Cosgrove notes that interpretations of photo 22727 have generally been framed by two "related discourses": what he terms the "one-world" discourse on the one hand and the "whole-earth" discourse on the other (262–263). The one-world discourse, he says, is concerned with ideas of communication and interconnectedness, but focuses more on the "global surface [. . .]. It is a universalist, progressive, and mobile discourse [. . .]. Consistently associated with technological advance, it yields an implicitly imperial spatiality, connecting the ends of the earth to privileged hubs and centers of control" (263). In contrast to this more imperialistic, disembodied view, the viewer may also come to understand the image as representing "the globe's organic unity" and "rootedness," in accord with what he terms the "whole-earth" discourse, which "emphasizes the fragility and vulnerability of a corporeal earth and responsibility for its care. It can generate apocalyptic anxiety about the end of life on this planet or warm sentiments of association, community, and attachment" (262–263). To fully oppose the one-earth discourse to the whole-earth discourse, however, is to fail to recognize the middle-ground between the two, and the ways in which each fosters different representations of connectivity. Moreover, appropriations of images of the earth contain variations that may be read in terms of both the one-world discourse (as more totalizing and universalizing, signifying networked communication and globalization) and the whole-earth discourse (as more inclusive and rooted, signifying local knowledge and individual accountability).[9]

While photo 22727 has come to be associated with both discourses, its initial reception was more readily associated with the social contexts of the emerging environmental movement of the United States in the 1970s. Its continued appropriation by environmental groups affords it a strong association with environmentalism even today. As Cosgrove describes, the photo's "apparent absence of cultural signifiers has made it a favored icon for environmental and human-rights campaigners and those challenging Western humanism's long-held assumption of superiority in a hierarchy of life" (*Apollo's Eye* 261):

> [T]he image [. . .] radically destabilizes the cultural part of the conventional meaning of Earth. [. . .] [I]t is no longer regarded as primarily the 'home of Man.' Earth is viewed as having

an intrinsic life, even its own intelligence as a homeostatic system, and all of its different species accorded dignity equal to that of humans. Humanity is decentered, and by regarding humans as merely one among a multitude of species the cultural variety which is a distinctive feature of our species is suppressed. (Cosgrove, "New World Orders" 128–129)

The image may then invoke in the viewer a sense of responsibility and kinship as opposed to distance or disembodiment. Again, the growing discourses of environmental conservation in the United States in the early 1970s contributed to such interpretations of the photo.

Contextualizations and Appropriations of Photo 22727

The publication of Rachel Carson's *Silent Spring* in 1962 is widely understood as one of the primary catalysts for contemporary environmentalism in the United States. Subsequently, in the decades following publication of *Silent Spring,* the public witnessed the steadily growing momentum of the environmental movement. While photo 22727 has the sort of staying power that has enabled its iconic status among environmental groups and with the public generally, it is important to recognize that it is just one of many rhetorical artifacts and objects of discourse associated with the emergence of environmentalism during that period. A general understanding of the social and political contexts surrounding the emergence of the photo helps to situate its rhetorical power and the associations it both reflected and perpetuated.

Just two years prior to circulation of photo 22727 on April 22, 1970, for example, the United States held its first Earth Day celebration, an event spearheaded by Democratic Senator Gaylord Nelson of Wisconsin. In 1971, just one year later, polls showed that "25 percent of the U.S. public declared protecting the environment to be an important goal, a 2,500 percent increase over 1969" ("Earth Day History"). Thus, according to Senator Nelson, "Earth Day launched the Environmental decade with a bang" ("Earth Day '70"). Soon after, photo 22727 became appropriated as the logo for subsequent Earth Day celebrations in the United States. Other appropriations of the photo include its use by the environmental group Friends of the Earth "to convey a message of global dwelling, care, and fragility," as well as its continual use in "antinuclear, environmental, and animal-rights campaigns" (Cosgrove, *Apollo's Eye* 263). In many ways, the photo has become a metonym for environmentalism. As Hariman and Lucaites describe, to view the image of the earth as a metonym for environmentalism would involve a "reduction of a more general construct," such as environmentalism, "to a specific embodiment," such as the photo of the earth, or the "Blue Marble" (89). As they describe, "[s]uch compositions have to be simultaneously personal and impersonal. [. . .] They depend on a thorough-going realism, but they motivate action in response to the general condition being represented rather than to the specific event of the picture" (89). Photo 22727 fits the bill well in this regard; the image of the earth portrays a convincing realism through its sharpness, its color, and the familiarity of the landforms; its small scale and circular shape also make it easily recognizable as an image of the earth. But it is not necessarily this realism that stirs the emotions; in fact, the photo's realism helps convey a distancing effect, or a sort of disembodiment that speaks more so to the one-world discourse. Rather, it is the condition being represented more generally, the implicit beauty and fragility of the planet, conveyed also by the informal naming "Blue Marble," that sparks feelings of personal responsibility and allows viewers to integrate their individual perspectives with their interpretation of the image. Viewers then employ their own experiences and understandings in their interpretation of the image, though these understandings are inextricably linked to the social and political contexts in which the image was presented. That is, photo 22727 first circulated in 1972, at a point when unprecedented acts of environmental legislation contributed to the growing discourses of environmentalism. In 1970, for example, the National Environmental Policy Act and the Clean Air Act were passed, Congress authorized creation of the Environmental Protection Agency, and the National Resources Defense Council was created. In 1971, the Animal Welfare Act was passed. In 1972, Congress passed the Federal Water Pollution Control Act, the Coastal Zone Management Act, the Ocean Dumping Act, and the Marine Mammal Protection Act. In 1973, Congress passed the Endangered Species Act (Kovarik). Outfitted with knowledge of new environmental legislation and organizations, the public's understanding of these discursive contexts likely helped shape their interpretation of the photo at that point in time. Conversely, the photo helped "communicate social knowledge [. . .] by tap[ping] into the tacit knowledge held by the audience as they are members of society" (Hariman and Lucaites 10).

In addition to understanding the image largely in terms of the "whole-earth" discourse around which it has been interpreted by its viewing publics, Cosgrove also acknowledges the vantage point of the Apollo 17 crew, who first witnessed the view that eventually became photo 22727: "Those few humans who actually witnessed the revolving terracqueous globe and who produced photo 22727 describe their experience in terms of awe, mystery and humility. The axis of world order, if it existed for them, stretched infinitely above and below the global surface" ("New World Orders" 130). This description of the astronauts' experiences in first viewing the earth from space not only reinforces understandings of mapping as a relational process but also helps to bring before our eyes a version of the image that is surreal and almost spiritual in nature. Cosgrove wants us to imagine the image of earth through the astronauts' eyes, invoking a sort of ekphrasis that transports us to that moment of witnessing prior to the photo's having been captured with the camera. Understanding the photo not only through the public lens of the whole-earth discourse but also from the vantage point of its producers affords yet an additional way of seeing that takes into account the astronauts' embodied experiences at a specific cultural moment, one that precedes even the production of the artifact itself.

As Hayles has described, experiences of embodiment are always enmeshed within a culture; they are contextual and linked to experiences of the world around us. To understand artifacts of cartographic practice and visual rhetoric as embodied knowledge allows us to consider the ways in which visual artifacts help provide more intimate understandings of or connections to a place, and in doing so, perhaps a closer relationship and feeling of responsibility toward it. An analysis of photo 22727 begins to show that to better understand the consequences of the rhetorical work of visual and material artifacts in the world, we must understand visual rhetorics as also attentive to studies of space, place, and the body.

VISUAL CULTURE, SPACE, AND THE BODY: A MOVE TOWARD MATERIALITY

Understanding photo 22727 as an artifact of visual rhetoric clearly helps demonstrates Cosgrove's view that "geography's words and images have always had a certain power to construct as much as to reflect the orders which it represents" ("New World Orders" 130). That is, as Pickles and others have pointed out, artifacts like the map participate in situated practices of visuality and are part of a broader visual culture. Understanding the significance of the artifact beyond its immediate function in the rhetorical situation is integral to framing visual rhetoric as a project of inquiry engaged in the practices of visuality. In this way, Olson et al. define visuality as referring not only to "images or visual media but [to] the totality of practices, performances, and configurations of the visual" (xvi–xvii). Compatible with the discussion earlier in the introduction, Carolyn Handa describes visual culture "as a subfield of cultural studies [that] focuses on vision as a starting point for tracing the ways cultural meanings form" (377). Visual culture is again implicated in the study of visual rhetoric, which she defines more broadly "as a discipline that focuses on the visual elements that persuade, taking culture as just one element among many: culture, along with images, sounds, and space, work together rhetorically to convince an audience" (377). A holistic approach to the study of visual rhetorics must then attend not only to these visual elements of persuasion but also to the situated and often multimodal practices in which they are immersed.

Also acknowledging the connections between visual culture and spatiality, Irit Rogoff writes that to open up "the field of vision as an arena in which cultural meanings get constituted, also simultaneously anchors it to an entire range of analyses and interpretations of the audio, the spatial, and of the psychic dynamics of spectatorship" (381). This understanding of spatiality as implicated in visual culture is likewise of interest to Handa, who writes: "If space, as Rogoff argues, is part of the intertextual mix that needs to be studied, we can learn much from those who critique, imagine, dictate, and analyze how space is inhabited" (378). Prelli too notes that built structures and places are "disposed rhetorically in their physical design so that their arrangement works to dispose the attitudes, feelings, and conduct of those who visit, dwell within, or otherwise encounter them" (13). Spatially-based, rhetorical artifacts such as maps, places, and built structures then tap into and rely on visual culture, which Kathryn Henderson further defines as "a way of seeing that reflects and contributes to the specific manner in which one renders the world," or "a particular way of seeing the world that is linked to *explicit material experience*" (197–198, emphasis added).

In understanding ways of seeing as tied to explicit material or corporeal experience in the world, Henderson broaches the idea of how visual culture affects contextualized, bodily experience. Gregory Clark describes a similar idea when he explores Kenneth Burke's theory of identification through the lens of American tourism. Here, Clark examines the ways in which national identity is shaped by public experiences of symbolic landscapes. Clark demonstrates how "the rhetorical power of a national culture is wielded not only by public discourse, but also by *public experiences*" (4). He describes identification with a place as necessarily tied to personal experience, when he writes eloquently of Burke's eventual meditation "on the way one's sense of self and possibility are transformed by the wordless symbols that constitute the experience of being present in a place" (29). Finally, Greg Dickinson and Casey Malone Maugh more explicitly address the connections between visual and material rhetorics in their analysis of visual rhetoric, place, and the Wild Oates Market, when they write that "buildings, and the institutions they house do not simply respond to the contemporary through visuality, instead they draw on the fully embodied subject" (260). Dickinson and Maugh go on to write that "how a definition or theory of visual rhetoric should address materiality is a complex problem, a problem for which we have, at best, partially constructed solutions" (260).

In this acknowledgement of the need for visual rhetoric to address materiality and the embodied subject, Dickinson and Maugh take a crucial step in moving toward an embodied, visual-material rhetoric that is attendant to the impact of space and place on the body. For, as they note, to understand visual, multimodal representations and physical structures as concerned with more than their immediately apparent features—to understand them also as embodied—allows us "to locate our bodies in relation to other bodies in the world" (Dickinson and Maugh 272). This act of locating allows us to engage in a richer mode of rhetorical analysis that considers the broader consequences of the rhetorical situation—that allows us to understand visual-material rhetoric as a project of inquiry. As Dickinson and Maugh also note, to incorporate materiality into the study of visual rhetoric, or conversely, to incorporate visuality into the study of material rhetoric, can pose a challenge, for it requires once again a nuanced interpretive lens that can accommodate multiple ways of knowing and multiple sites of inquiry. It is thus one goal of this book to take up such a challenge.

To this end, I propose a methodological framework for understanding visual-material rhetorics that applies and extends Carole Blair's theory of material rhetoric and merges that theory with Michel Foucault's concept of heterotopias. Foucault's concept of heterotopias focuses on understanding spaces as heterogeneous, selective, contested, and culturally situated. Blair's theory of material rhetoric then helps us better understand the consequences of different spaces on the body. While Blair speaks primarily of physical, material spaces, I describe how we may extend her theory to also account for visual and multimodal spaces and artifacts. In the next chapter, I take a closer look at discussions related to material rhetoric, situating among them Blair's theory of material rhetoric and Foucault's concept of heterotopias. I understand the ideas of Blair and Foucault as stepping stones that allow for a point of entry into the more important question of how a visual-material rhetorical approach can provide a window into the larger consequences that these artifacts have in the world. I maintain that an approach that merges and extends their theories can provide the foundation for a visual-material rhetoric that not only accounts for the multimodal, spatially-situated artifact but is also mindful of its impact on the embodied subject. Again, it is the acknowledgement and understanding of embodiment that I feel begins to situate visual-material rhetorics as a continued project of inquiry as opposed to a more insular and immediately available analytical tool.

2 The Visual-Material Spectrum

To understand the study of visual-material rhetorics as a sustainable project of inquiry that can provide a window into the larger consequences that these artifacts have in the world means uncovering an analytical approach that can account more explicitly for the ways in which visual-material artifacts and particular spaces can shape or influence the practices of the contextualized body. In other words, visual-material rhetorics must account not only for the cultural work of the spatially-situated artifact but also specifically for its impact on the embodied subject. To understand rhetoric as embodied is to explore rhetorical practice as it manifests through the action of the body, or "to follow the expressive ebb and flow of expressive energy through human bodily activities: through gesture, through contact with and manipulation of objects, through movement and space" (Marback 62). This chapter not only takes a closer look at conversations focused more explicitly on the idea of embodiment and material rhetorics but also sees Blair's theory of material rhetoric and Foucault's concept of heterotopias as appropriately situated when considered among them. Subsequently, I contend that the theories of Blair and Foucault can function symbiotically to allow for a more nuanced and embodied understanding of visual-material rhetorics as a mode of inquiry.

A Brief Note about Turns and Spectrums

As we begin to consider the relationship between visual and material rhetorics, it would seem plausible to question the need for a "material turn" within visual rhetoric. On the one hand, to consider the relationship of material rhetoric to visual rhetoric as constituting a "turn" seems almost obligatory, given the tendency of the humanities and social sciences to mark new conversations and disciplinary foci as such. On the other hand, to consider the notion of a "material turn" is to potentially misrepresent what I contend is a more integrated, already existing relationship between the visual and the material. W.J.T. Mitchell's "pictorial turn," [1] for example, helps us understand that it is possible to view a potentially increased focus on material rhetorics not as a full turn, per se, but rather as an already present component of visual rhetoric. For Mitchell, embodied knowledge and materiality are implicitly accounted for in the pictorial turn and its attendant modes of visual practice. He notes that however the pictorial turn may be defined, it is fundamentally a "postlinguistic, postsemiotic rediscovery of the picture as a complex interplay between visuality, apparatus, institutions, discourse, bodies, and figurality" (16). Here, he describes the picture not as an artifact in and of itself but rather as implicated in broader cultural, relational, and physical contexts.

The analysis of photo 22727 in chapter one, for example, described the iconic image as situated in the discourses of the emerging environmental movement of the United States in the 1970s. When understood in light of the whole-earth discourse, the image may be read as prompting a more embodied experience that fosters a sense of personal responsibility toward the earth. To understand artifacts of visual rhetoric and cartographic representation as implicated in the pictorial turn is to view them as fostering embodied knowledge, as inviting a more intimate understanding of or connection to a particular place. We may see, then, that Mitchell's acknowledgement of these interactions may indeed be read as accounting for materiality and embodied knowledge as a component of the visual—as part of the shift that already constitutes the pictorial turn. When understood in this light, the inclusive nature of Mitchell's pictorial turn provides a useful starting point for envisioning the movement between the visual and the material as happening along a spectrum. I argue that to understand the relationship between the visual and material as such allows for a more inclusive mode of knowing that opens up rather than closes off interpretive possibilities—that accounts more readily for the movement between these modes and their interplay, such that we may more directly engage in the study of visual-material rhetorics as embodied knowledge and a sustainable project of inquiry.

MATERIALITY, SPACE, AND THE BODY

As described briefly in the introduction, a theory of material rhetoric, as conceptualized most clearly in Carole Blair's 1999 study of five

U.S. memorial sites, has at its core a focus on the impact of spatially-situated texts on contextualized, bodily experience. A closer look at recent work related to material rhetoric reveals that studies in this area may be understood as situated along a continuum. At one end of this continuum are analyses with a primary focus on physical space and a subsequent focus on the impact of those spaces on the bodies residing within them; at the other end of the continuum are analyses with a focus on the body first and foremost, and a secondary, contextualizing focus on the sociocultural contexts which make possible such analyses of the body. And of course, there are those analyses that are situated not along one end or the other, but someplace in the middle. In this chapter, I examine recent conversations explicitly related to studies of material rhetoric; included among them are the theories of Blair and Foucault, which I see as an integral component of any discussion and subsequent theory of visual-material rhetorics.

The earlier work of Carole Blair contains visible hints of what would later become the theory of material rhetoric set forth in her essay "U.S. Memorial Sites." In an earlier essay, for example, Blair and her co-authors, Jeppeson and Pucci, describe the rhetorical impact of the Vietnam Veteran's Memorial, suggesting that its visibility has been enhanced through its reproduction in popular culture—ideas that then receive further treatment in "U.S. Memorial Sites" ("Public Memorializing" 263). While Blair and her co-authors do not yet use the term "material rhetoric," their criteria for postmodern architecture are not incompatible with the goals of a material rhetoric. These criteria primarily focus on a memorial's "melding of incompatible symbols, forms, styles, and textures within a particular structure," and its integration of regional or historical characteristics and forms (267). Deborah Fausch takes further the idea of a postmodern architecture, forwarding the idea of a feminist architecture that is quite similar to how we might understand material rhetoric. An architect herself, Fausch feels that feminist architecture may be designated as such "if it fostered an awareness of and posited a value to the experience of the concrete, the sensual, the bodily—if it used the body as a necessary instrument in absorbing the content of the experience" (42). This idea too is compatible with the general goals of Blair's later theory of material rhetoric, as outlined in the essay "Contemporary U.S. Memorial Sites," in which she understands the action of the body as a necessary component in absorbing even a portion of a site's meaning.

As described in the Introduction, L.J. Nicoletti incorporates what may be read as a Blairian framework in her creation of an assignment geared toward helping her first-year writing students cope with the events of September 11th and respond to the types of memorialization they were witnessing in the mass media. Again, while this is not necessarily a book about public memory, Nicoletti aptly points out that to uncover the arguments built into commemorative sculptures or monuments may also make us more attuned to the ideological agendas often perpetuated through particular renderings or portrayals of politicized events (53). Similarly, Barbara Biesecker writes that "claiming and representing the past is far from being an innocent affair" ("Remembering" 168). As Biesecker's important analysis of the Women in Military Service for America Memorial (WIMS) demonstrates, a memorial's ostensible goals may differ from its more subtle rhetorical work. On the surface, for example, the WIMS appeared to acknowledge the millions of women who have served in the U.S. military since the Revolutionary War (Biesecker 165). As a critical analysis of the memorial reveals, however, the memorial's revisionist history served to alienate the women whom it was intended to valorize. Within the WIMS, an exhibit gallery that should have promoted the accomplishments of individual women soldiers functions more generally to "mark the regular rhythms and daily practices of our nation's service women" (166). As a result, she writes, "typicality rather than rarity subtends the order of things" (166). Consequently, these artifacts have the effect of problematically implying a representative, universalizing narrative, perpetuating a "seemingly complete, unabridged history of women in the U.S. armed services" (166). Instead, Biesecker calls for something akin to a multimodal rhetorical approach characterized more so by its ability to foster embodied knowledge. A material rhetoric approach indeed makes room for the contextualized nuances of multimodal, embodied experience that influence the cultural moments in which we interact with rhetorical artifacts.

Barbara Dickson and Dan Brouwer each consider the ways in which material and visual rhetorics function within the contexts of the mass media and the public, though their objects of analysis are less concerned with public memory and national identity than with the processes that make possible specific representations of the body in popular culture. Brouwer examines the sociocultural contexts that inform the practice of wearing HIV/AIDS tattoos. Considered a form of "self-stigmatization," the practice of wearing the tattoo is "a particular communicative and performative strategy grounded in visibility politics and practiced in the context of AIDS activism" (Brouwer 206). Noting its precariousness as a social act, Brouwer writes that the tattoo "simultaneously disrupts expectations of the appearance of health and challenges 'norms' of patient behavior, yet [. . .] also invites surveillance [. . .] and runs the risk of reducing the wearer's identity to 'disease carrier'" (206). Brouwer's rich analysis not only illuminates the motivations behind this powerful social practice but also provides a more empowered and "sensitive understanding of the communication practices of marginal or stigmatized social groups" by understanding how "performative communication" illustrates the connections "between the margins and center of power" (217–218). Also focused on the sociocultural contexts that allow for resistance through bodily inscription, Dickson analyzes the iconic 1991 *Vanity Fair* cover photo of actor Demi Moore, which depicted her "seven months pregnant and wearing nothing but diamonds" (297). Dickson questions whether the representation of Moore's pregnant body can be seen as "liberat[ing] the feminine body," as Moore claimed it did in her description of the photo as a "feminist statement" (297). Dickson understands the photo as a "textual event" and considers the cultural contexts that shape its "production and reception" (299). A material rhetorical analysis of these "bodily, visual, and textual" inscriptions allows Dickson to better understand the hegemonic discourses that inform bodily inscription and how such inscriptions get instantiated materially (311–312).

Building on Blair and Dickson's approaches and giving more equal treatment to spatial analysis and the impact of physical space on the body, Mary Lay Schuster's material rhetorical analysis of Baby Haven, a free-standing birth center in middle America, describes the consequences of the center on the minds and bodies of clients who come there seeking an "ideal birth," or one that resists "the construction of their pregnant bodies as risky entities best managed by medical experts" (3). Schuster describes Baby Haven as a rhetorically powerful space that allows clients to "rewrite cultural inscriptions" that construct the body, in order to forward an understanding of the birthing process that works against the hegemonic biomedical model (30). Also affording more equal treatment to physical space and its impact on bodies, but discussing in addition the material component of textual artifacts, Christina Haas analyzes the work of a Permanent Injunction posted on the front door of an Ohio abortion clinic. The Injunction is meant to deter abortion protesters and create a safe space for women who seek to have an abortion performed (Crowley 359). Haas describes the Injunction in terms of its material rhetorical and cultural dimensions, but frames the document more so as a mediating device that helps to make tangible the "conceptual distinction between public and private" (234). Through its articulation of spatial boundaries that clarify where protests may be staged, the Injunction acts on the bodies within and outside of the abortion clinic to protect the employees of the clinic, thus fostering a better sense of safety among clinic workers (224).

Multimodality as a Component of Materiality

In addition to material rhetoric's focus on physical space and the built environment, fields related to composition and media studies have begun to acknowledge the materiality of multimodal texts and digital artifacts. Hayles, for example, argues for a multimodal, material literacy in her keynote address at the 2002 Computers and Writing Conference. While the recent focus on visual rhetoric is a clear step in the right direction, she says, "we need to develop modes of critical attention responsive to the full range of [. . .] signifying elements in electronic work, including animation, sound, graphics, screen design, and navigational functionalities" ("Deeper" 371). In acknowledging that our vocabulary for analyzing printed text is insufficient for the critique of digital texts, Hayles broaches the intersections of digital texts and materiality (373). To this end, she first notes that electronic texts require a "critical language" sensitive to the interplay of word and image; she then sees this interplay as indicative of larger issues related to the broader practices of multimodality: "This new critical vocabulary," she says, "will further realize that navigation, animation, and other digital effects are not neutral devices but designed practices that enter deeply into the work's structures; it will eschew the print-centric assumption that a literacy work is an abstract verbal construction and focus on the materiality of the medium" (373). Aligned with Hayles, Barbara Warnick notes that "the material form of a representation is an intrinsic dimension of the user's experience of it, and so critical approaches need to take into account the materiality of the text, as well as its content and style of expression" (328). The GPS, as I will describe in chapter four, is one example of a multimodal, material artifact that not only epitomizes the variety of content, contexts, and styles of expression that the text may produce but also helps illuminate the value of a visual-material rhetorical approach for the study of multimodal artifacts.[2]

Similar to the discussions of material rhetoric that I've described here, this book focuses on analyses of specific spaces, the artifacts that contribute to the rhetorical power of those spaces, and consequently the impact of those rhetorical spaces on the bodies that inhabit or once inhabited them. Again, this book also moves across visual-material artifacts such as spatial representations of mill life, park memorials, and maps, in order to more overtly call attention to or reconcile versions of contested space and show the value of visual-material rhetorics within and beyond the field of rhetoric. Through a consistent focus on these artifacts' selectivity; their material, visual, and textual composition; and their subsequent impact on human, posthuman, and non-human bodies residing or once residing in the spaces they represent, this book understands these sites as visual-material rhetorics of heterotopic space. To arrive at such an understanding then requires that we acknowledge the ideas of Blair and Foucault as part and parcel of a theory of visual-material rhetorics.

In Foucault's Theory of Discourse, an Understanding of Space as Rhetorical (Or, En Route to a Visual-Material Rhetorics of Heterotopic Space)

While Foucault is of course no stranger to scholars of rhetoric, his concept of heterotopias is not one that is frequently invoked within the field; rather, it is more common, especially among new graduate students of rhetoric, to read Foucault's work within the context of his theory of discourse. Because his work on heterotopias focuses more directly on his theory of spatiality, it is often more familiar to those who study geography or critical cartographies. Nonetheless, Foucault's theory of discourse and his theory of space share some common themes; in fact, the curious reader may notice the seeming stylistic and thematic similarities between these two areas of his work. Moreover, readers may account for these similarities by noting the close chronological proximity of his initial articulations of these ideas. That is, the ideas underpinning Foucault's essay, "Of Other Spaces," preceded publication of *The Archaeology of Knowledge* by only two years. While

"Of Other Spaces" was officially published in 1984 as "Des Espaces Autres" in the French journal *Architecture-Mouvement-Continuité*, it was in March of 1967, in France, that Foucault first gave the lecture that would then serve as the basis for this essay, in which he posits his theory of heterotopias (22). Only two years later, in 1969, *The Archaeology of Knowledge*, or *L'Archeologie du Savoir*, in which Foucault sets out much of his theory of discourse, was first published in France. Subsequently, to read these two works side by side—especially to read "Of Other Spaces" alongside "The Unities of Discourse" within *The Archaeology of Knowledge*—quite seemingly invites a reading of space as discursive. As Patricia Bizzell and Bruce Herzberg write in *The Rhetorical Tradition,* Foucault's theory of discourse

> [seeks to] restore to discourse its character as an event. [. . .] [It] describes the relationship between language and knowledge; the functions of disciplines, institutions, and other discourse communities; the ways that particular statements come to have truth value; the constraints on the production of discourse about objects of knowledge; the effects of discursive practices on social action; and the uses of discourse to exercise power. (1127)

Likewise, Foucault's theory of space may be understood as an active endeavor—one that is concerned with teasing out the relationships between space and knowledge, understanding how spaces may constrain meaning by appearing simple while also concealing knowledge, and understanding how "our life is still governed by a certain number of oppositions that remain inviolable, that our institutions and practices have not yet dared to break down" ("Of Other Spaces" 23).

Foucault's theory of heterotopias asks that we take a close look at our hierarchic "history of space," which he notes may be traced roughly to the Middle Ages—we must understand this hierarchic ensemble of places in order to expose the different relationships that delineate them—this "ensemble of places," he says, includes "sacred places and profane places; protected places and open, exposed places; urban places and rural places. [. . .] It was this complete hierarchy, this opposition, this intersection of places that constituted what could very roughly be called medieval space: the space of emplacement" ("Of Other Spaces" 22). Today, he says, this space of emplacement "has been substituted for extension," which is defined by "relations of proximity between points or elements; formally, we can describe these relations as series, trees, or grids"; the intricacies of these sites then manifest, he says, when dealing with "the storage of data or of the intermediate results of a calculation in the memory of a machine; the circulation of discrete elements with a random output (automobile traffic is a simple case, or indeed the sounds on a telephone line). [. . .] In a still more concrete manner, the problem of siting or placement arises for mankind in terms of demography" (23). Interestingly, we see in Foucault's description of the contemporary manifestation of the hierarchic ensemble of places, or in the transition from emplacement to extension, allusions to the spatial dimensions and problematics of mediated bodies functioning within a technologically mediated society. Moreover, the "relations of proximity," or the series, trees, and grids to which Foucault refers, then make possible human practices that, as Biesecker might put it, work both "within and against the grain" to resist hegemonic constructions of space (357). That is, as Biesecker describes within the context of discussing the "implications of Foucault's work for Rhetoric" (352), the practices that are made possible through these grids also "carry within themselves what Foucault calls 'a kind of virtual break' out of which transgression may ensue" (356). Such acts of transgression thus constitute a form of resistance. This notion of resistance, Biesecker feels, is rooted in Foucault's "non-monumentalized conception of power" (354). I argue here that Biesecker's ideas about Foucault's theory of resistance are not only relevant to the field of rhetoric in general but also to how we might understand the rhetorical study of space more specifically.

Biesecker suggests that Foucault indeed has a theory of resistance, and that it is embedded largely in his understanding (and our *mis*understanding) of power (*pouvoir*). That is, "to understand power only as oppressive is reductive" (Biesecker 354). Rather, when we understand that the meaning of the French verb *pouvoir* loses some of its dimension in the English translation, we begin to understand the ways in which power can be productive for Foucault. Here, Biesecker quotes from Gayatri Spivak: "*Pouvoir* is of course 'power.' But there is also a sense of 'can-do'-ness in *pouvoir* [. . .] it is the commonest way of saying 'can' in the French language" (qtd. in Biesecker 355). Power then conveys not just limits, but also a "being-able" that happens at multiple levels of practice that at once rely on "existing lines of sense" and "carry within themselves a 'virtual break'" (357). Resistance thus works "within and against the grain" (357). It "names the nonlegible practices that are performed within the weave but are asymmetrical to it. As Foucault put it, 'They are the odd term in relations of power'" (357).

Foucault's program in "The Unities of Discourse" is compatible with Biesecker's description of how resistance works "within and against the grain" (357). Moreover, "The Unities of Discourse" not only puts into clearer context how scholars of rhetoric might proceed in thinking about discourse but also how they might begin to understand a Foucauldian notion of space. Like his theory of heterotopias, Foucault's theory of discourse critiques historiography and its propensity toward creating normative continuities; it does so by problematizing a "whole mass of notions" such as tradition, origin, influence, causality, unity, development, and coherence (*Archaeology* 21), all of which have the effect of "master[ing] time through a perpetually reversible relation between an origin and a term that are never given, but always at work" (22). Foucault's notion of heterotopic space is founded on a similar idea that contrasts "indefinitely accumulating time" with "time in its most fleeting, transitory, precarious aspect" in order to problematize normative continuities and tease apart the various realities that compose a given space; these realities are time sensitive, reliant on cultural contexts, and often oppose or challenge others' claims to knowledge (Foucault, "Of Other Spaces" 26).

In the "Unities of Discourse," Foucault asks us to understand the rules such that we might break them, so to speak; that is, we need to "accept the groupings that history suggests only to subject them at once to interrogation; to break them up and then to see whether they can be legitimately reformed; or whether other groupings should be made; to replace them in a more general space which, while dissipating their apparent familiarity, makes it possible to construct a theory of them" (*Archaelogy* 26). Foucault's theory of space also implicitly asks us to interrogate the groupings that history suggests and to question whether they may be viewed from the point of their discontinuity rather than from that of their perceived continuity. He asks that we resist hierarchies of space and look instead for knowledge in unfamiliar places; in doing so, he not only asks us to understand space as epistemic but also as discursive. And if we subscribe to Bizzell and Herzberg's reading of Foucault's notion of discourse as rhetorical, then we may likewise understand space not only as discursive but also as rhetorical. That is, Bizzell and Herzberg note that although

> Foucault avoids talking about rhetoric, preferring *discourse* as his comprehensive term, there is no question that his theory addresses a number of ideas that are central to modern rhetoric. He makes a powerful argument that discourse (for which we may read *rhetoric*) is epistemic; he states in compelling terms that discourse is a form of social action; he enriches and complicates the notion of context with a network of archives, disciplines, institutions, and social practices that control the production of discourse. (1128)

Foucault's theory of space and its implicit discursivities help illuminate the power structures inherent in spatial relationships such that we might then find within them the sort of "virtual break" that Biesecker describes. In this way, as Biesecker puts it, "power names not the imposition of a limit that constrains human thought and action but a being-able that is made possible by a grid of intelligibility" (356). This sort of "being-able," however, is nonetheless constrained in terms of its implications for the fully embodied, individualized subject. On the one hand, as Biesecker describes in depth, Foucault's later work (particularly, she says, in *The Uses of Pleasure*), addresses "the 'stylized practices of the self' or 'aesthetics of existence' [which] may be read as a concerted effort on his part to specify the place and function of the deliberate intending subject whose acts, though made possible by the social apparatus or field, cannot be reduced to the mere playing out of a code" (358). On the other hand, however, subjectivity, for Foucault, even in his later work, is still understood as a consequence of societally imposed power relations, even though he sees subject positions as uniquely individual. (Biesecker 360). In this way, for Foucault, human beings may be understood as actively taking part in the environments in which they are situated, but these acts of participation are imposed on them by cultural norms and societal groups.[3] This negotiation between the practices of the self and the imposing cultural constructs that influence those practices has been the site of much contestation over what, for Foucault, has been viewed as constituting his theory of resistance.

FOUCAULT, HETEROTOPIAS, AND MATERIAL RHETORIC

In describing the universalizing practices of space such that those practices provide opportunities for the identification of breaks or fissures, or possibilities for resistance, Foucault does take us closer to the notion of an embodied subject who may work with and against the grain of societal constructs. Nonetheless, Foucault's view of the universalized, resisting subject as always already responding to imposed societal constructs may be understood as constraining or limiting possibilities for a more empowered view of embodied knowledge. Such limited possibilities for the universalized body arguably perpetuate what Hayles has termed Foucault's "erasure of embodiment" (194). I argue here that while Foucault's theory of heterotopias allows us to identify the universalizing functions of space such that we might simultaneously identify ways to work against them, his theory does not readily address to a fuller extent the idea of individual, bodily experience within heterotopic spaces. In other words, I argue that to expose the normative functions of space and the possibilities they provide for resistance is not necessarily to understand or describe those functions and possibilities as implicated in embodied, material practice. To derive such insights about embodied, spatial rhetorics from his theory of space thus requires extending Foucault's work to account more explicitly for the embodied nature of physical space. This is precisely the point at which Blair's work becomes useful.

That is, while Foucault helps us to understand space as rhetorical, some trickiness arises when we must also understand space and its attendant visual and material artifacts not only as rhetorical and powerful in that "can-do" sense but also as embodied. To help make these connections clearer and to address the dilemma of embodiment requires several steps. First, it is necessary to describe how heterotopias may be specifically characterized according to Foucault's theory. Next, it is necessary to explain the limitation of Foucault's work, primarily as articulated through Hayles' notion of his erasure of embodiment, which, while compatible with an understanding of his theory of resistance, calls for a more explicit focus on the idea of "how embodied humans interact with the material conditions in which they are placed" (Hayles 195). At this point, then, it becomes possible to describe how Blair's approach works to supplement or extend the potential limitation of Foucault's theory by providing a point of entry into a visual-material rhetorics of heterotopic space.

Defining Heterotopias

Foucault suggests that we all reside in heterogeneous spaces—that "we do not live in a kind of void, inside of which we could place individuals and things"; rather, "we live inside a set of relations that delineates sites which are irreducible to one another and absolutely not superimposable on one another" ("Of Other Spaces" 23). We can try to characterize these sites by "looking for the set of relations by which a given site can be defined" (23). Sites of transportation, for example, may include streets or trains; "sites of temporary relaxation" may include "cafes, cinemas, [or] beaches" (24). Foucault is most interested, though, in those sites "that have the curious property of being in relation with all the other sites, but in such a way as to suspect, neutralize, or invert the set of relations that they happen to designate, mirror, or reflect" (24). Such spaces, he says, fall into two main categories: utopias and heterotopias.

Utopias, he says, "are sites with no real place"; they present a perfected vision of society and are "fundamentally unreal" (24). Heterotopias, rather, may be found in "every culture, in every civilization" (24). Heterotopias, Foucault writes, are "real" [4] places "that do exist and that are formed in the very founding of society" (24). They are akin to "counter-sites," or an

> effectively enacted utopia in which the real sites, all the other real sites that can be found within the culture, are simultaneously represented, contested, and inverted. Places of this kind are outside of all places, even though it may be possible to indicate their location in reality. Because these places are absolutely different from all the sites that they reflect and speak about, I shall call them, by way of contrast to utopias, heterotopias. (24)

Foucault's notion of the heterotopia as counter-site, representation, reflection, and discursive, contested site would seem to open up the possibility for resistance, or for a site that, as Biesecker has put it, "names the nonlegible practices that are performed within the weave but are asymmetrical to it" (357). As the analysis of photo 22727 in chapter one helped demonstrate, a map is indeed a representation of a particular "real" territory, though often conveys multiple ideas about a place. These multiple ideas about a place are often borne out of knowledge claims that result in competing or contested discourses about what counts as the most "accurate" representation of a single territory. Memorials too count as places that represent, contest, speak of, or invert the "real" sites that they call out or commemorate. Thus, maps as well as commemorative artifacts may be understood as heterotopias. Already, then, it is possible to see that Foucault's terminology can account for the varied contexts that help shape our potential understandings of a place. Following his initial defining of heterotopias, Foucault then provides some criteria to help identify these sites and the characteristics that qualify them as such.

Foucault's Six Principles of Heterotopology

In his characteristic style, Foucault embarks on what he terms a "systematic description" of heterotopias. He refers to this description as "a sort of simultaneously mythic and real contestation of the space in which we live," and thus terms it a "heterotopology," which then consists of six main principles (24). While his categories for describing heterotopias are useful in and of themselves, I also use them as heuristics to consider as heterotopic such artifacts as cartographic texts and public commemorative sites. In the discussion that follows, I describe Foucault's six principles of heterotopology.

Principle One. Foucault says that heterotopias can be found across cultures—in all cultures, in fact. No "universal form" of heterotopia exists; while all heterotopias vary in their makeup, however, they may be classified "into two main categories," which he calls "crisis heterotopias" and "heterotopias of deviation" ("Of Other Spaces" 24). Crisis heterotopias are "privileged," "sacred," or "forbidden places reserved for individuals who are in a state of crisis," such as, he suggests, "adolescents, menstruating women, pregnant women, the elderly, etc." (24). Crisis heterotopias, he says, may function as boarding schools, places of military service, honeymoon hotels, or places without geographical markers (24–25). Foucault also describes heterotopias of deviation: "those in which individuals whose behavior is deviant in relation to the required mean or norm are placed" (25). On the one hand, heterotopias are geographically diverse and culturally specific—there is no universal heterotopia. On the other hand, heterotopic spaces of crisis and deviation function as disciplinary mechanisms that house deviant, universalized, disciplined bodies. Thus, it bears noting that Foucault's emphasis here is on the ability of the site itself to house the universalized or hidden body rather than on the interaction of the individual-ized, corporeal body with the material environment.

Principle Two. Heterotopias specific to a given society can change or shift over time as that society's history develops (Foucault, "Of Other Spaces" 25). That is, "the same heterotopia can, according to the synchrony of the culture in which it occurs, have one function or another" (25). Foucault cites the cemetery as his main example here, noting the changes it has undergone since the eighteenth century, relative to shifting societal norms regarding burial of the dead and the physical location of the cemetery within or outside of the city.[5] The main purpose of Foucault's second principle of heterotopology is to describe a society as malleable, as capable of changing a dominant heterotopia or imposing new heterotopic models to which cultures and social groups then respond.

Principle Three. Next, Foucault describes the heterotopia as "capable of juxtaposing in a single real place several spaces, several sites that are in themselves incompatible" or contradictory, such as the theater, the cinema, or the garden" ("Of Other Spaces" 26). He cites the garden as the oldest example of the heterotopia and describes its history as a sacred space: it is "the smallest parcel of the world and then it is the totality of the world"; it "has been a sort of happy, universalizing heterotopia since the beginnings of antiquity" (26). Interesting here is his notion of the garden as a universalizing heterotopia. The garden, that is, in Foucault's description, gains its power and recognition largely through its incorporation into a hierarchic ensemble of places; in doing so, however, its universalizing function also potentially distracts us from understanding how bodies might engage or interact within it. To better understand the more specific ways in which these and other sites participate in versions of lived experience again requires that we build upon or extend Foucault's work to account for how bodies engage and are engaged by their material environment.

Principle Four. Foucault writes, "Heterotopias are most often linked to slices in time," a notion he terms *heterochronies.* He cites museums and libraries as heterotopias of "indefinitely accumulating time"; in museums and libraries, he says, "time never stops building up and topping its own summit" ("Of Other Spaces" 26). Foucault does not describe or account for bodily practice and human interaction as epistemic components of the timelessness of these spaces; however, we might consider the role of embodied experience and the practices of memory in the heterochronous properties of such sites as the museum and the library. In contrast to these heterotopias of "indefinitely accumulating time," Foucault also notes that some heterotopias may be linked to "time in its most fleeting, transitory, precarious aspect, to time in the mode of the festival," citing as examples fairgrounds and vacation villages (26). These heterotopias, he says, are "not oriented toward the eternal, they are rather absolutely temporal" (26). Here Foucault expresses a view similar to that which he describes in the "Unities of Discourse"—one which understands time not as perpetuating an origin narrative but as working against normative continuities in its focus on always shifting temporalities. This movement between timelessness (or "indefinitely accumulating time") and timeliness ("time in its most fleeting, transitory, precarious aspect") also resists the linear narrative and works against the grain to identify "the nonlegible practices that are performed within the weave but are asymmetrical to it" (Biesecker 357). As Biesecker has discussed, the rhetorical work of memory texts are often an "effect of what and how we remember, and the uses to which those memories are put" (168–169). And as Marback has stated, an artifact may subsist throughout time, but its "meaning and significance are in the moment" (52–53).[6]

In addition to memory texts that function in an explicitly commemorative capacity, mapped representations and other multimodal rhetorical artifacts may participate in memory practices that do other types of knowledge-making. As chapter five will address in more depth, maps may be produced in response to temporally-specific events and may serve as timeless artifacts that represent a continuous, ongoing struggle to advocate for the bodies represented through them. Brooke also considers the function of memory practices in digital environments when he notes that the "externalization of memory has become an accepted and even integral part of society" (786). In this way, digital mapping devices such as the GPS, for instance (as chapter four will discuss in more detail), may participate in memory practices through their ability to store and retrieve addresses and destinations, thus mediating bodily experiences of remembering that have the capacity to change the ways that humans interact with and understand their environments. From the perspective of visual-material rhetorics, then, the significance of memory practices, understood as bearing diverse possibilities for rhetorical work, lies in their potential to engage with and advocate for the bodies represented through their spatial and textual artifacts.

Principle Five. Foucault also suggests that "[h]eterotopias always presuppose a system of opening and closing that both isolates them and makes them penetrable" ("Of Other Places" 26). These sites, he says, tend not to be "freely accessible like a public place. Either the entry is compulsory, as in the case of entering a barracks or a prison, or else the individual has to submit to rites and purifications" (26). Other entries may appear "pure and simple," he says, but may "hide curious exclusions" (26). Entry here is but an illusion. He cites the motel room where a person might go with their mistress, where sex is both "sheltered" and "hidden, kept isolated [. . .] without being allowed out in the open" (27). Interesting here is the notion of the hidden exclusion, or that of the unsaid, to employ language more closely associated with Foucault's theory of discourse. In the "Unities of Discourse," for example, Foucault recommends that we "reconstitute another discourse, rediscover the silent murmuring, the inexhaustible speech that animates from within the voice that one hears, re-establish the tiny, invisible text that runs between and sometimes collides with them" (*Archaeology* 27). Thus, the question posed in the "analysis of thought [. . .] is unfailingly: what was being said in what was said?" (27–28). This call to action may be applied to an analysis of spatiality as well; that is, the system of opening and closing presupposed by heterotopias implicitly leaves room to identify and resist these curious exclusions.

Principle Six. Lastly, Foucault's sixth principle of heterotopology states that all heterotopias "have a function in relation to all the space that remains. This function unfolds between two poles," which take the form of 1) heterotopias of illusion or 2) heterotopias of compensation:

> Either their role is to create a space of illusion that exposes every real space, all the sites inside of which human life is partitioned, as still more illusory. [. . .] Or else, on the contrary, their role is to create a space that is other, another real space, as perfect, as meticulous, as well arranged as ours is messy, ill constructed, and jumbled. This latter type would be the

heterotopia, not of illusion, but of compensation. ("Of Other Places" 27)

Of heterotopias of illusion, Foucault takes the brothel as his example, though curiously he does not provide substantive discussion in this regard. It is possible to infer, however, the connotation of the brothel as a hidden space that exposes the realities of behavior deemed societally deviant and makes no apologies for the actions carried out within it. Instead, he devotes more time to the notion of heterotopias of compensation, which he says were epitomized during the "first wave of colonization in the seventeenth century," both in the English Puritan societies and the South American Jesuit colonies (27). In these societies, "existence was regulated at every turn" (27). Of heterotopias of compensation, Foucault writes:

> The daily life of individuals was regulated not by the whistle, but by the bell. Everyone was awakened at the same time, everyone began work at the same time; meals were at noon and five o'clock; then came bedtime, and at midnight came what was called the marital wake-up, that is, at the chime of the churchbell, each person carried out her/his duty. (27)

Interestingly, while Foucault describes heterotopias of compensation here in terms of their impact on the body, he does so not through discussion of the individualized body but rather through reference to universalized, docile, disciplined bodies. As Hayles notes in her critique of Foucault's "erasure of embodiment" when speaking of the work of the Panopticon, while the "bodies of the disciplined do not disappear in Foucault's account, the specificities of their corporealities fade into the technology as well, becoming a universalized body worked upon in a uniform way by surveillance techniques and practices" (*Posthuman* 194). On one level, then, we might read into Foucault's heterotopia of illusion, a power to recognize the resistance inherent in the unfiltered realities that happen within a deviant space; on another level, perhaps Foucault's sixth principle of heterotopology best exemplifies the limitations of his theory of heterotopias through its overt focus on the universalized, hidden, disciplined body.

In the Apparent Limitations of Foucault,
Room for Enriched Understanding

This book employs and extends Foucault's six principles of heteroto-pology in order to account for the heterotopic qualities of artifacts such as the mill, the memorial, and the cartographic text—all objects of visual-material rhetorics. Foucault's theory helps us to see that hetero-topias are concerned with temporally-situated representations of con-tested space that may not be reduced to monologic discourses; instead, as Foucault describes, the spaces in which we live are heterogeneous and irreducible to one another—they are multi-layered and discursive. Nevertheless, while Foucault's theory describes the particular places that themselves may constitute heterotopias, his work, at least in de-tailing this concept, does not directly address how such spaces act on or engage the body. That is, his work does not seem to account for the particularities of bodily practice within specific environments; rather, he focuses more so on the properties and characteristics of a particular space. Given the seeming limitations of this understanding, it is poten-tially left to the reader to presume that these spaces may indeed engage the bodies inhabited by them and imagine how they might do so—unless, of course, we uncover a means of accounting for this gap. It is within the space of this gap that Hayles' important critique of what she refers to as Foucault's "erasure of embodiment" becomes relevant (*Posthuman* 194). Moreover, I suggest that this limitation also provides an opportunity to take a closer look at how we might supplement and employ Blair's framework to account for visual-material rhetorics as embodied heterotopic space.

Looking to Foucault's work in *Discipline and Punish,* Hayles notes that we can see evidence for Foucault's erasure of embodiment in his acknowledgement that "the Panopticon was never built" (*Posthuman* 194). As Foucault writes in *Discipline and Punish,* "The fact that [the Panopticon] should have given rise, even in our own time, to so many variations, projected or realized, is evidence of the imaginary inten-sity that it has possessed for almost two hundred years" (205). And as Hayles points out, Foucault argues that the Panopticon "'must not be understood as a dream building; it is the diagram of a mechanism of power reduced to its ideal form; its functioning, abstracted from any obstacle, resistance or friction, must be represented as a pure architec-tural and optical system; it is in fact a figure of political technology that may and must be detached from any specific use'" (qtd. in *Posthu-*

man 194). Hayles notes that while the "abstraction of the Panopticon [. . .] into a system of disciplines dispersed throughout society gives Foucault's analysis its power and universality," it also draws our focus away from understanding how individual, physical, contexualized bodies "impose, incorporate, and resist incorporation of the material practices he describes" (194). This line of critique highlights the point of contention between Foucault's acknowledgment of the practices of self as described by Biesecker, and Hayles' critique of Foucault. That is, Foucault describes seeing the subject as actively participating in practices and patterns that "'he finds in his culture and which are proposed, suggested, and imposed on him by his culture, his society and his social group'" (Foucault qtd. in Biesecker 360). In this view, culture and societal groups may be read as imposing the practices to which the subject then responds, resists, or makes visible. Such an understanding is also compatible with Foucault's second principle of heterotopology, which, as mentioned earlier, describes society as malleable, as capable of imposing new heterotopias to which cultures and social groups then respond. Certainly, here, the power of cultures and social groups convey not just limits, but also the "being-able" that happens at multiple levels of practice that at once rely on "existing lines of sense" and "carry within themselves the 'virtual break,' as Biesecker describes (357). Not incompatible with this view and building on it, Hayles asks how we might turn the tables a bit, or how we might understand contextualized, situated bodies as more explicitly doing the imposing and proposing—as incorporating or resisting the practices imposed upon them through cultures and social groups. The distinction is a subtle but important one for understanding how it is that a theory of visual-material rhetorics may be seen as advocating for a more empowered view of the interaction between bodies, spaces, and material discursive practice. Implicitly consistent with this mode of thinking, for example, Brouwer's analysis of the practice of wearing an HIV/AIDs tattoo may then be viewed as a communicative performance that resists the dominant hegemonic discourse of HIV/AIDS. While the choice to wear the tattoo is nonetheless a response to the particular sociocultural contexts that make the choice an option to be pursued in the first place, Brouwer's more empowered emphasis is on the nature and implications of that choice "in the context of AIDS activism" (Brouwer 206). Hayles then goes on to write that part of what gives the Panopticon its force is the fact that it hides the specific power of the disciplinarian, abstracting that power "into a universal, disembodied gaze" and hiding "the limitations of corporeality" (*Posthuman* 194). The individual bodies disciplined through the Panopticon are also hidden and universalized (194). When scenarios that involve "embodied agents are considered, limits appear that are obscured when the Panopticon is considered only as an abstract mechanism" (194). In not recognizing these limits, Hayles argues that Foucault is complicit in "the Panoptic move of disembodiment" and ultimately posits "a body constituted through discursive formations and material practices that erase the contextual enactments embodiment always entails" (*Posthuman* 194).

Largely as a result of this important work on the Panopticon, many scholars have critiqued the subsequent "universality of the Foucaultian body," which may be read as an outcome of his focus on discourse as opposed to embodiment—a focus that "makes it difficult to understand exactly how certain practices spread through a society" (Hayles, *Posthuman* 195). Thus, as Hayles rightly notes, "[b]uilding on Foucault's work while going beyond it requires understanding how embodiment moves in conjunction with inscription, technology, and ideology. Attentive to discursive constructions, such an analysis would also examine how embodied humans interact with the material conditions in which they are placed" (195). Blair's theory of material rhetoric provides a good point of entry for building on Foucault's work and going beyond it to better understand the consequences of bodily interaction within specific spaces. Where Foucault first helps us identify and characterize heterotopic sites and the means by which they expose other such spaces, Blair helps us talk about the consequences of heterotopias on bodily practice. Blair's material rhetoric works well to describe the ways in which spatially-focused artifacts such as maps, industrial sites, or green spaces, for example, bear consequence on the body and thus constitute visual-material rhetorics of heterotopic space. I therefore suggest that Foucault's theory of heterotopias is better understood as more fully contributing to a rhetorical framework for understanding spatially-based artifacts when supplemented with or understood within the context of Blair's work on material rhetoric.

A Closer Look at Blair's Material Rhetoric

Blair's approach to material rhetorical analysis provides a toolkit for analyzing visual-material texts. As mentioned in the introduction, she begins from the point of acknowledging that within the field of rhetoric, "we lack an idiom for referencing talk, writing, or even in-scribed stone as material"—that we struggle with "the lack of a ma-terialist language about discourse" (17). She poses five questions that help to redefine what counts as a text; in doing so, she works toward a theory of material rhetoric that can account for the broader impacts and consequences that spatially-situated texts can have in the world, beyond the site of their immediate function in the rhetorical situation. To this end, she asks: "(1) What is the significance of the text's mate-rial existence? (2) What are the apparatuses and degrees of durability of displayed by the text? (3) What are the text's modes or possibilities of reproduction or preservation? (4) What does the text do to (or with, or against) other texts? (5) How does the text act on people?" (30). Within the context of each of these questions, Blair also poses addi-tional sub-questions for analysis. In the context of asking how the text works with or against other texts, for example, she notes some further possibilities for exploration, such as a text's potential for "enabling, appropriating, contextualizing, supplementing, correcting, challeng-ing, competing, or silencing" (39). By using as her toolset a group of questions that address the more nuanced aspects of a text's durabil-ity, modes of reproduction, visibility, and materiality, Blair accesses a materialist approach to discourse that has been under-realized within rhetorical analysis. Blair's approach creates a solid framework for mate-rial rhetorical analysis that, as I will show, may also be used as heuristic to understand the text as a visual-material heterotopic space.

For Blair, creating a space for material rhetoric means moving be-yond the traditional view of rhetoric as symbolic (concerned primarily with meaning) and immaterial (or fleeting and ephemeral); to this end, she says:

> No text *is* a text, nor does it have meaning, influence, political stance, or legibility, in the absence of material form. Rhetoric is not rhetoric until it is uttered, written, or otherwise manifested and given presence. Thus, we might hypothesize as a starting point for theorizing rhetoric that at least one of its basic char-acteristics (if not the *most* basic) is its materiality. (18)

Blair does not suggest, however, that we must altogether stop theoriz-ing rhetoric as symbolic; instead, she rightly points out that symbolic-ity needn't necessarily be conceived of as ephemeral.[7] And as Blair goes on to explain, symbols are, in and of themselves, material, non-ephemeral components of a language system. They ultimately refer us outside of themselves, directing us to their referents for meaning, asking us "to treat that meaning as if it were the real dimension of rhetoric, or at least the most important one" (19). In other words, it is possible to focus on the idea of rhetoric's symbolicity or even the ways in which symbols are themselves material, but even so, "it is problematic to treat rhetoric as if it were exclusively or essentially symbolic" (19). This is because, as Blair points out, "there are some things that rhetoric's symbolicity simply cannot account for" (19). One of these things, she says, is its potential for consequence (19). Too little emphasis, Blair maintains, is placed on a text's greater consequence; moreover, when the idea of consequence "is addressed at all, it is typically advanced as a reason to study the construction (production values, if you will) of a particular text; and it is frequently understood narrowly as 'success' or goal fulfillment" (21–22). To focus less on a text's production can bring us to the point of more meaningful analysis—one that is focused more on the text's embodied, contextualized interaction with the world. Blair then asks us to step outside of our analytic comfort zone; rather than ask "what a text means," she says, "[i]f rhetoric's materiality is not a function of its symbolic constructions of meaning, then we must look elsewhere: we must ask not just what a text means but, more generally, what it does; and we must not understand what it does as strictly adhering to what it was supposed to do" (23). Blair argues that while "everyone seems to know that rhetoric is not exclusively about production, and more specifically, that it has consequences that exceed goal fulfillment [. . .] hardly anyone seems willing to address it as anything else" (22). This "anything else," she says, is rooted in materiality.

Rhetoric then functions in the realm of the material world—of the lived experience, and as such, it acts on the body and has a varied capacity for consequence. Already, we may begin to see how a study of heterotopic spaces that also understands those spaces as materially rhetorical can work against the "erasure of embodiment" that Hayles critiques. For example, we may understand the factories at the Lowell Mills to be physical spaces that, together, constituted a particular work environment for the Mill Girls who labored there in the 1800s. Foucault might categorize these spaces as crisis heterotopias. A material rhetorical analysis of these crisis heterotopias would consider the ways in which they form environments that ultimately bear consequence on the bodies that inhabit them. Thus, a space may qualify as a heterotopia, and the heterotopia subsequently may be understood as rhetorical, or as an object of material rhetoric. In this sense, we may conceive of rhetoric as perhaps a component of heterotopias, or by contrast, we may conceive of heterotopias as rhetorical.

VISUAL-MATERIAL RHETORICS OF HETEROTOPIC SPACE: LOOKING AHEAD

Heterotopias represent the varied, contested spaces in which we live. In conceiving of heterotopic spaces as reflecting the various contexts that shape and influence bodily experience, we may begin to understand space as bearing the potential for consequence. Thus we may begin to understand heterotopic space as rhetorical—as visual-material textual sites that, when understood also through the lens of Blair's material rhetoric, have the potential for consequence and can influence lived experience. In this view, conversely, we may also understand rhetoric as concerned with space and place. With these ideas in mind, it becomes possible to arrive at an initial understanding of a visual-material rhetoric of heterotopic space.

In the next chapter, I test these ideas through the lens of the first of the three rhetorical contexts underpinning the organization of this book. That is, one goal of this book is to better understand the value of visual-material rhetorics from three interconnected perspectives: that of their influence on the body, the posthuman body, and the non-human body. Chapter three accounts for the impact of visual-material rhetorics on contextualized, bodily experience by analyzing the maps, way finding devices, green spaces, and public commemorative sculptures at the Lowell Mills National Historical Park in Lowell, Massachusetts. The chapter not only demonstrates how the park's spatially-focused, cultural artifacts engage visitors and facilitate their navigational decision-making, but also how these artifacts reflect and perform the impact of the mills on the lives of the Mill Girls who labored there in the early 1800s. With these ideas in mind, I demonstrate how a visual-material rhetorical approach can help us engage with greater empathy in the lives and struggles of tangential or underrepresented groups (such as, in this case, the lives of the Mill Girls), and create room for more nuanced and empathetic understanding of the contexts in which these women lived and worked.

3 Empathizing with Marginalized Bodies

In the early 1820s, Boston capitalists, led by Francis Cabbot Lowell, purchased ownership rights to land and water along the Merrimack River in East Chelmsford, Massachusetts and began building what would soon become home to the Lowell Mills (Josephson 11; Dublin 99). The early 1800s brought with it the spread of industrialization and subsequently the expansion of New England's textile industry (Josephson 204–5). And as the textile industry grew in New England, so did the textile mills of Lowell: "Additional mills were constructed until, by 1840, ten textile corporations with thirty-two mills valued at more than ten million dollars lined the banks of the river and nearby canals" (Dublin 99). Lowell, however, was not proximally close to any neighboring residential areas from which workers could commute, and thus the Boston Associates, who oversaw the mills, "constructed row upon row of corporation-owned brick boardinghouses next to each mill complex" (Stanton 48). These boardinghouses eventually became home to most of the eight thousand women who worked at the mills (Dublin 99).

Employees of the early Lowell Mills were generally young women with puritan values who came from nearby farming communities in rural Massachusetts. Enticed by promises of better wages, safe and pleasant room and board, and opportunities for educational growth, women traveled from surrounding rural communities to experience urban life and earn a relatively high salary that they could then share with their families. The decision by the Boston Associates to recruit young women to work in the mills was ostensibly a way of working to prevent the sort of "poverty and misery that the English Industrial Revolution had brought to British workers and towns" (Stanton 48). With these goals in mind, Lowell's founders "sought to avoid the most degrading human and social effects of industrialization by hiring republican-spirited young women who would work in the mills for only a few years" (48). During this time, these young women would live in the boardinghouses, due not only to the practicality of their proximity to the mills but also "because of the belief of the times that young single women needed close moral supervision and chaperonage" (48). The Boston Associates worked to perpetuate the idea that "economic and technological progress could coexist within the purifying natural landscapes" that have, as Cathy Stanton notes, long characterized American democratic ideals (48). As such, the mill owners created tidy and aesthetically pleasing images of mill life that were reinforced by landscapes consisting of tree-lined parks and "walkways along the canal system [. . .] Conveying the ideas of historian Robert Dalzell, Stanton writes that "this spatial harmony was designed to produce social harmony" (48–49). Thus, the large investments in corporate housing, coupled with the high turnaround of mill workers and the carefully cultivated landscapes were aimed at "'controlling the process of change, of channeling its direction so that it did not sweep everything before it'" (Dalzell qtd. in Stanton 49). And, as Stanton says, "for a very short time, the plan worked" (49).

In the view of feminist geographer Doreen Massey, "the identities of place are always unfixed, contested, and multiple"; any given place should not be viewed "by placing boundaries around it," but rather, by considering the ways in which that place is implicated in the "mix of links and interconnections" to the places beyond it (5). Likewise, Lowell was interconnected with the growth of New England's textile industry and soon expanded its mills in both size and function; subsequently, the lives of the women working in the mills were implicated in this shift. The social experiment that characterized the early years of the Lowell Mills then began to lose its idyllic quality. Lowell soon became the largest manufacturer of textiles in the nation, which greatly impacted the lives of the mill workers. As textile production increased in the surrounding region, competition among mills also resulted in overproduction and drove down the price of finished cloth. Profits soon declined and likewise led to declining conditions for the mill workers. These declining conditions included reduced wages and increased expectations for worker productivity (Dublin 100). The mill operatives did not easily accept these conditions, and between 1834 and 1836, "they went on strike to protest wage cuts, and between 1843 and 1848 they mounted petition campaigns aimed at reducing the hours of labor in the mills" (Dublin 100). The Ten Hour Movement was the most well-known of these campaigns and was most active in 1845. The campaign was aimed at reducing the factory workday from 14 hours to 10 hours per day, and was headed by Sarah Bagley, herself a former mill worker and contributor to the *Lowell Offering*.

The *Lowell Offering* was a literary magazine that was published from 1840 to 1845, amidst wage cuts, protests over long workdays, and deteriorating conditions at the mills. Overseen by the mill owners and the clergymen of Lowell, this literary magazine was written by the mill operatives and first published under the editorship of Harriet Farley (Foner 19). The magazine ostensibly was intended to provide a forum in which the mill workers could talk about their lives in the mills and write poetry and prose that would provide an outlet for their creative and cultural expression. However, the tone and content of these narratives often contradicted one another and may be read as reflecting an underlying tension between the attempts at freedom of expression on the part of the female operatives and the attempts of the mill owners to censor this expression.[1] From these narratives, then, it is fair to discern that the Mill Girls often struggled with the conditions of their work environment and the dissonance they experienced when leaving their rural homes for urban industrial life.

In this chapter, I travel to the original site of the Lowell Mills, to explore what, in 1978, became designated as the Lowell Mills National Historical Park (LMNHP) (Stanton 3). I explore the contemporary visual-material rhetoric of the park and question whether its spatial layout engages visitors' bodies in a manner similar to how the original mills impacted the bodies of the Mill Girls, or whether the LMNHP has at all revisioned the lives of the Mill Girls. Such questions regarding consistency and revisioning help speak to issues around the significance of the contemporary text's material existence, its "modes or possibilities of reproduction or preservation," and how these texts work "with, or against" historic accounts—all questions addressed in Blair's theory of material rhetoric (30). These questions are also grounded in an understanding of the early mills as heterotopic space.

As discussed in chapter two, heterotopic spaces are heterogeneous sites that "always presuppose a system of opening and closing that both isolates them and makes them penetrable" (Foucault, "Of Other Spaces" 26). Commemorative sites or green spaces, for example, often require that we tap into existing cultural norms and symbols in order to infer their meaning. They are also selective, providing points of entry into understanding certain aspects of the environment while excluding other features of the terrain; they may, as Foucault says, appear "pure and simple," but may also "hide curious exclusions" (26). Representations of space, such as the LMNHP map, also invoke specific cartographic conventions such as color, projection, scale, and the use of lines and numbers in order to represent terrain in a particular way. These modes of knowing not only tap into existing cartographic design conventions specific to certain groups, but also make use (to varying degrees) of audience, context, and purpose. Foucault's notion of heterotopias, then, when understood also in terms of its potential for considering embodiment, is highly related to the notion of context— of being able to understand the complex set of relationships within and outside of a particular space or representation of space, which can then help account for the corporeal impact and rhetoricality of that space. These contexts require that we have at least some schema for understanding the complex ontologies of the site itself in order to understand its consequences on the bodies residing within it. Artifacts at the LMNHP, for example, describe a correlation between the increasingly oppressive work environments in the Lowell Mills in the 1800s and the simultaneous expansion of New England's textile industry. That is, industrial growth created competition between local mills; consequently, a need then arose for increased levels of production at the individual mills—a responsibility that fell largely on the shoulders of the female operatives who worked there. In this chapter, then, I am interested in exploring the value of a visual-material rhetorical approach for illuminating the ways in which the discursive nature of the park's spatial layout and its subsequent impact on visitors can help foster a more empathetic understanding and imagining of early mill life.

ON CHAPTER ORGANIZATION AND DATA COLLECTION

To better understand the value of visual-material rhetorics in illuminating the work of the park's portrayal of mill life, I focus on four specific spaces and representations of space within the LMNHP, all of which may be understood as pertaining to the lives of the Mill Girls. First, I examine the park map that appears in the brochure given to visitors. I argue that through its aesthetics, graphical features, and selectivity, the map makes claims to knowledge about which features are to be deemed most important. I then describe the park's layout and the implications of its being built into the landscape of downtown Lowell. Next, I focus on two smaller parks within the LMNHP itself: Lucy Larcom Park and Boardinghouse Park and the sculptures within them. I argue that the sculptures within and spatial layout of these smaller parks and green spaces reflect and perform the crisis heterotopias and heterotopias of compensation that characterized the early Lowell Mills. For example, we may understand the factories at the Lowell Mills to be physical spaces that, together, constitute a work environment that was often perceived as oppressive and taxing to the Mill Girls. Foucault might categorize these spaces as crisis heterotopias, given the common dilemma shared by the Mill Girls and instantiated through the physical spaces of the mills. These spaces were also portrayed as neat, tidy, and well-regulated, and better allowed mill owners to control and regulate the activities and schedules of the Mill Girls. Understood as heterotopias of compensation, such spaces reflect a societal order in which existence is "regulated at every turn" (Foucault, "Of Other Spaces" 27). Lastly, I analyze the public commemorative sculpture, *Homage to Women,* which is situated just behind the visitor center and is included by many of the park's tour guides in their walking tours. This sculpture too acts on the body, allowing its visitors to arrive at various interpretations, many of which generally suggest that the memorial communicates the Mill Girls' dissatisfaction for their working and living conditions and their desire for improved circumstances. The spatial layout of the LMNHP as a whole also communicates the difficulty of the Mill Girls' work environment and the constant surveillance under which they lived.

Because the park focuses on so many rich and diverse aspects of Lowell's industrial heritage and its role in the Industrial Revolution, it was necessary to delimit my analysis to these four main features of the park. The park also works to promote and describe the cultural heritage of the different groups that immigrated to the United States during the 19th and 20th centuries, and uses a combination of museum displays, walking tours, guided tours, and canal tours to do so (Stanton xiii). While my main focus of interest is on the women who labored in the mills during the antebellum (pre-Civil War) period, it is necessary to note that the description of the Mill Girls and their place in Lowell's history is but one facet of the park's work. I thus delimit my analysis to those park features and green spaces that best demonstrate the mill's impacts on the minds and bodies of the Mill Girls, their subsequent influence on park visitors, and the ways in which a visual-material rhetorical approach can help illuminate those impacts.

Throughout my analysis, I also consider whether or how these experiences at all mirror the Mill Girls' impressions of the mills as conveyed in the *Lowell Offering.* A close reading [2] of their early writings reveals that many women frequently described feeling overwhelmed by the tall brick buildings, or were made to feel dizzy by the sheer volume of bodies in the boardinghouses and dining halls. Many mill operatives also expressed a "disillusionment with factory life" and a longing for home (Montrie 21). The writings of these women often "centered on differences in landscape and were coupled with romantic musing about the beauty and spiritual fullness of nature" (Montrie 21). As a result, many of these narratives include geographically-rich descriptions of the spaces inhabited by the Mill Girls and clear descriptions of the impact of these spaces on their bodies. Because the Mill Girls' narratives remain available in library archives, I feel it is important and useful to be able to understand the LMNHP also within the context of the firsthand experiences of the women who once worked and resided at the mills. While my reading of the Mill Girls' narratives in the *Lowell Offering* does inform my background knowledge on the topic, it does not constitute the bulk of my analysis; rather, I view this archival research as a necessary but ancillary component of my primary analyses. As such, I invoke quotes from these archival narratives or related secondary sources when they help to work with or against observations or interpretations about the park itself. In other words, my observations and analysis of the park itself drives my discussion.

It is also necessary to note that I do not speculate in this chapter about the intention of the artists or architects involved with various aspects of the park's design; rather, I focus on the consequences of these spaces on the visitor's bodily experience while in them—consequences that may or may not have been intended for by its designers. I did not attempt to interview the artists or architects involved with the design of certain green spaces at the LMNHP because, as Blair notes, the focus of material rhetoric ought not to be on the author's intent, per se. Instead, such analyses ought to focus on how a space is perceived by those who visit or inhabit it, the consequences of those perceptions on the body, and subsequently on the ways in which, through this bodily experience, we come to better understand the rhetorical situation.

Foucault's theory of heterotopias and Blair's theory of material rhetoric underpin my analyses of the park and its artifacts. I also incorporate my own observations of park visitors [3] and, where relevant, points from an interview conducted with a senior park ranger at the LMNHP. My main goal, however, has been to conduct what I consider to be a visual-material rhetorical analysis of specific green spaces and artifacts at the LMNHP. Thus in the analysis that follows, I demonstrate not only how spaces within the park aid in performing the crisis heterotopias and heterotopias of compensation of the early Lowell Mills, but also how the rhetoricality of these heterotopic spaces then allow park visitors to achieve greater understanding of the impacts of the mills on the minds and bodies of the Mill Girls.

THE BROCHURE MAP AND ENTERING THE PARK

The Lowell Mills park map (Figure 2) is distributed in conjunction with the National Park Service Department of the Interior. A paper version that is part of a larger brochure containing additional historical information is provided to all tourists who enter the park, and an abridged, downloadable version depicting only the park map is also available online. The map in Figure 2 depicts the online version of the map. The map functions as a detailed "locator map," or the type of cartographic genre that "describes a place or feature in relation to nearby places or features with which the viewer might be acquainted [. . .] or might want to visit" (Monmonier, *Mapping it Out* 76–77). Monmonier also notes that the locator map is a popular cartographic genre in "studies of architectural history, historical preservation, and industrial archaeology" (77). Given this description, the map of the LMNHP fits the category of locator map well and emphasizes Lowell as a national historical site.

The park map is widely available to visitors entering the visitor center. Any visitors who drive in and park their car in the main lot must obtain a parking permit at the visitor center, during which time a park ranger also provides the map. Even for visitors who do not drive in, or who choose to park elsewhere, these maps are difficult to miss upon entering the visitor center. While it is possible to enter the park at various points, the visitor center is clearly demarcated and serves as the park's main point of entry. From a Foucauldian perspective, the map makes strong knowledge claims about what information is to be privileged within the park and has great potential to direct the visitor's navigational experience.

As Pickles says, Foucault has shown us that "mappings always produce worlds by combining and recombining relations and ideas" (Foreword ix). And as Crampton and Krygier also note, Foucault's attentiveness to geography and spatiality makes him "of particular interest because he shows that many problems of politics require spatial knowledge" (14). As a representation of a particular "real" site that simultaneously represents, contests, or inverts other sites within the park, the LMNHP map makes knowledge claims and functions as a visual-material heterotopic text.

The map depicts the city of Lowell and the surrounding region (a small area encompassing roughly three square miles), from the University of Massachusetts Lowell campus and the Merrimack River to the north, to Centralville to the east, Pawtucketville to the west, and Route 495/the Lowell Connector to the south (the highway that many, if not most visitors will need to take when driving to the park). This larger vicinity is shown mainly to provide context for the location of the historic park, and so the map cannot be used as any real navigational tool beyond the park limits, as the areas outside of the park are shaded and include only sporadic street names and geographic landmarks.

The paper version of the map, included in the larger brochure, comprises a full side of the double-sided document and also contains the heading, "Visiting Lowell." The other side of the brochure provides a brief history of the mills with some illustrations and includes the following four sections: "Lowell and the Industrial Revolution," "Working at the Mills" (which also discusses the Mill Girls), "Immigrant Lowell," and a section called "Prosperity and Decline" (*Lowell National*). Both the downloadable map (shown in Figure 2) and the paper-based brochure map, however, send the message that to visit the LMNHP is indeed to visit the city of Lowell itself. This message is consistent with the park's immersion in the landscape of downtown Lowell, and as Stanton notes, many park rangers will explain on their tours that the park is inextricably linked with the city, "whose life today is in many senses an outgrowth of its industrial past [. . . and] to get this idea across, the rangers sometimes use a phrase that was heard often during the park's early years: 'The park is the city and the city is the park'" (3).

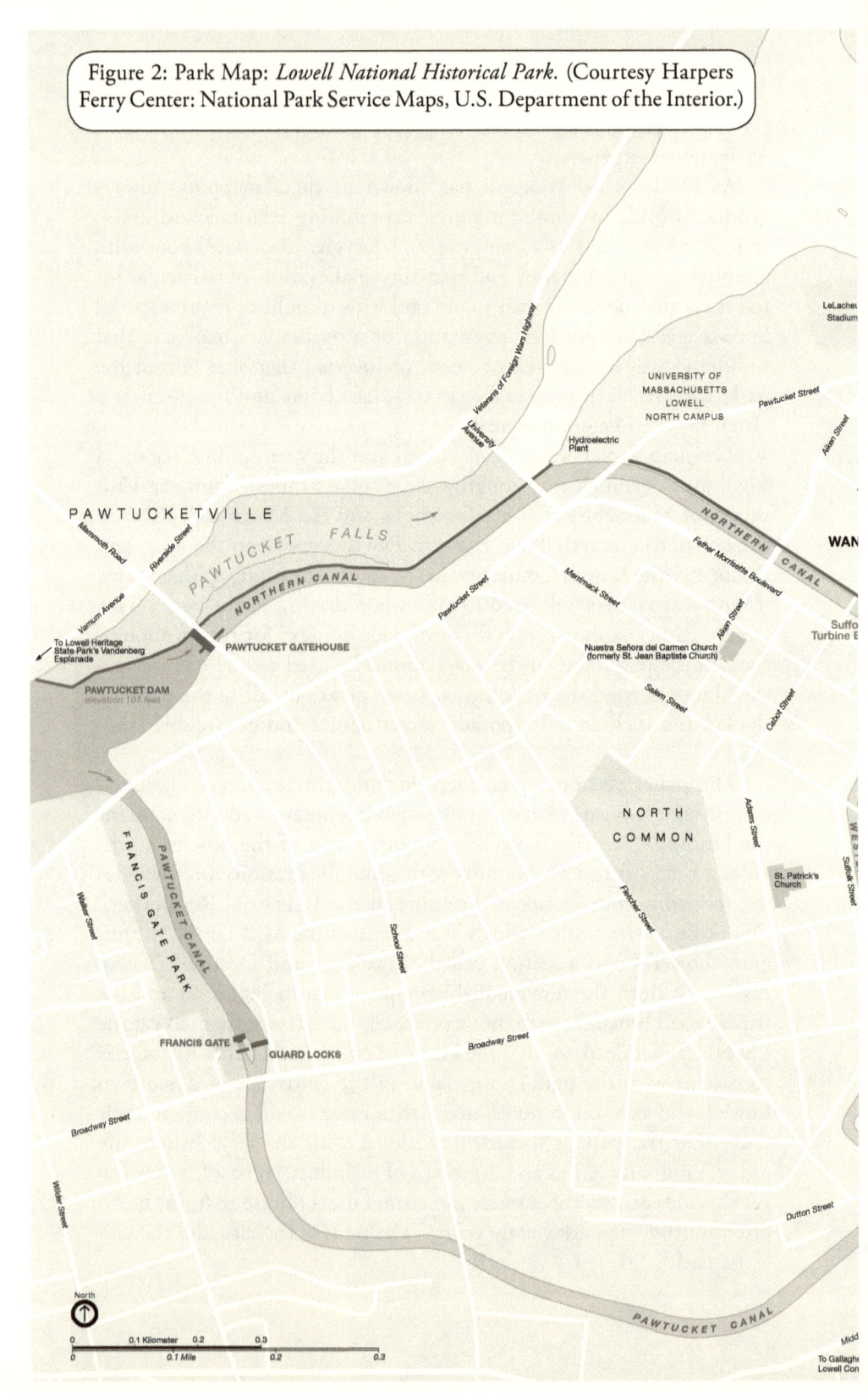

Figure 2: Park Map: *Lowell National Historical Park.* (Courtesy Harpers Ferry Center: National Park Service Maps, U.S. Department of the Interior.)

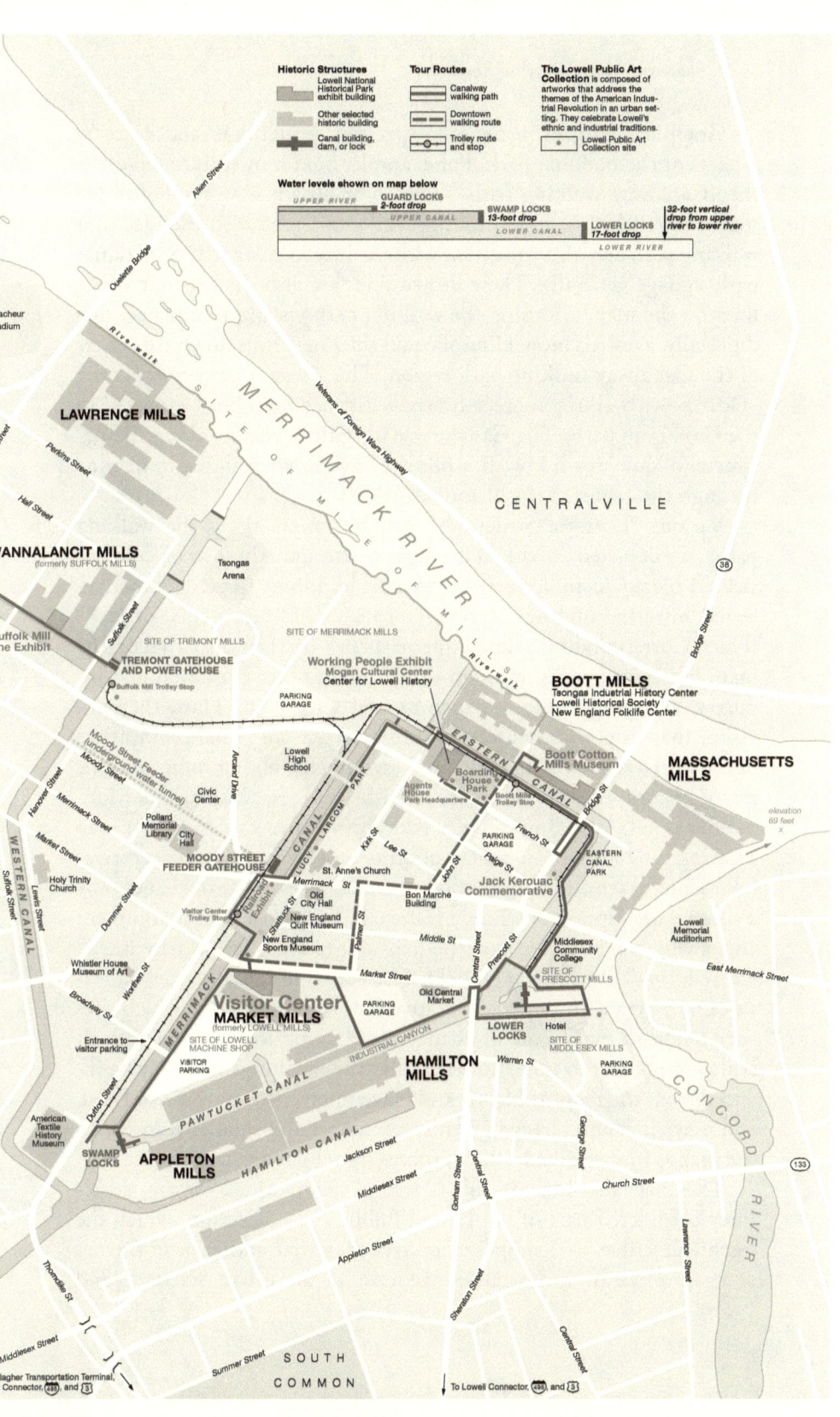

Historic Structures
Lowell National Historical Park exhibit building
Other selected historic building
Canal building, dam, or lock

Tour Routes
Canalway walking path
Downtown walking route
Trolley route and stop

The Lowell Public Art Collection is composed of artworks that address the themes of the American Industrial Revolution in an urban setting. They celebrate Lowell's ethnic and industrial traditions.
Lowell Public Art Collection site

Water levels shown on map below

UPPER RIVER	GUARD LOCKS 2-foot drop			32-foot vertical drop from upper river to lower river
	UPPER CANAL	SWAMP LOCKS 13-foot drop		
		LOWER CANAL	LOWER LOCKS 17-foot drop	
				LOWER RIVER

LAWRENCE MILLS

CENTRALVILLE

MERRIMACK RIVER
SITE OF MILE OF MILLS

Riverwalk

Ouellette Bridge

Aiken Street

Perkins Street

Hall Street

WANNALANCIT MILLS
(formerly SUFFOLK MILLS)

Tsongas Arena

Suffolk Street

Suffolk Mill Line Exhibit

SITE OF TREMONT MILLS

SITE OF MERRIMACK MILLS

Riverwalk

Veterans of Foreign Wars Highway

36

Bridge Street

TREMONT GATEHOUSE AND POWER HOUSE
Suffolk Mill Trolley Stop

Working People Exhibit
Mogan Cultural Center
Center for Lowell History

BOOTT MILLS
Tsongas Industrial History Center
Lowell Historical Society
New England Folklife Center

MASSACHUSETTS MILLS

PARKING GARAGE

Moody Street Feeder (underground water tunnel)
Moody Street

Accord Drive

Lowell High School

EASTERN CANAL

Boott Cotton Mills Museum

elevation 69 feet

Hanover Street

Merrimack Street

Market Street

Civic Drive

Pollard Memorial Library

City Hall

MOODY STREET FEEDER GATEHOUSE

Agents House
Park Headquarters

Boarding House Park

Boott Mills Trolley Stop

French St

PARKING GARAGE

EASTERN CANAL PARK

Lowell Memorial Auditorium

Holy Trinity Church

Dummer Street

St. Anne's Church

Merrimack St
Old City Hall

Kirk St

Lee St

Palmer St

John St

Jack Kerouac Commemorative

Bridge St

Dutton Street

Lowell Public Art Collection

Whistler House Museum of Art

Worthen St

Broadway St

Visitor Center Trolley Stop

Railroad Exhibit

Shattuck St

LUCY LARCOM PARK

New England Quilt Museum

New England Sports Museum

Bon Marche Building

Middle St

Central Street

Prescott St

Middlesex Community College

SITE OF PRESCOTT MILLS

East Merrimack Street

MERRIMACK

Visitor Center
MARKET MILLS
(formerly LOWELL MILLS)
SITE OF LOWELL MACHINE SHOP

Entrance to visitor parking

VISITOR PARKING

PARKING GARAGE

Old Central Market

INDUSTRIAL CANYON

HAMILTON MILLS

Market Street

Warren St

LOWER LOCKS

Hotel
SITE OF MIDDLESEX MILLS

PARKING GARAGE

CONCORD RIVER

American Textile History Museum

SWAMP LOCKS

APPLETON MILLS

PAWTUCKET CANAL

HAMILTON CANAL

Jackson Street

Middlesex Street

Appleton Street

Central Street

Gorham Street

Church Street

George Street

133

Thorndike St

WESTERN CANAL

Suffolk Street

Lewis Street

Warren St

Shenton Street

Lawrence Street

Clinton Street

CONCORD RIVER

Middlesex Street

SOUTH COMMON

Summer Street

To Lowell Connector, 495, and 3

Gallagher Transportation Terminal, Lowell Connector, 495, 3

Graphical features in the map specifically highlight and describe aspects of the national park. For example, bold brown lines represent the "Canalway walking path." The color brown may symbolize associations with dirt, earth, hiking—all consistent with the idea of a walking path and also consistent with the brown color of U.S. national park signage generally. These brown lines are also thicker than other lines on the map, affording the walking path visual prominence. Additionally, a dotted brown line of equal thickness runs down the center of the Canalway walking path region. The dotted line represents the "Downtown walking route," shown as distinct from but connected to the Canalway path. The Canalway path guides visitors along the perimeter of downtown Lowell, while the Downtown walking route runs through the center of the historic district.

Various "Historic Structures" can be found along the walking paths, as depicted in the map's legend through the use of colored, shaded blocks to indicate either exhibit buildings (green); "other selected historic buildings" (gray); or canal buildings, dams, or locks (blue). Corresponding colored square shapes are then displayed on the map, indicating parks, museums dedicated to transportation and industry, and other buildings such as Old City Hall. Thus, the park visitor may infer a causal relationship between the visual prominence of the Canalway walking path and Downtown walking route, and the many historic sites to which these paths lead. In this way, as Monmonier notes, "[a]dding causally relevant features can convert a map showing only location to a map offering an explanation or interpretation. Careful selection of causally related features can provide not only a concise description of where a feature is, but also a cogent argument about why it is where it is, or perhaps of greater interest, why it isn't where it isn't" (*Mapping it Out* 78). These sorts of causally relevant features may create an implicit narrative for the visitor about which sites they ought to make note of as they travel along the walking path. It is also important to note, however, that the meanings that may be inferred from the map, as Harris and Hazen note, are not static or fixed, and may shift and change given the contexts and moments in which the map is immersed and the purposes for which it is used (53).

The map also includes in its legend designations for any structures considered part of the Lowell Public Art Collection, which the legend describes as "composed of artworks that address the themes of the American Industrial Revolution in an urban setting," and "celebrat[ing] Lowell's ethnic and industrial traditions" (*Lowell National*). These artifacts are indicated through the placement of a small brown dot, and include sculptures within sites such as Lucy Larcom Park and Boardinghouse Park, as well as the *Homage to Women* sculpture. While these sculptures do have formal names of their own, their titles are not individually or specifically listed on the map, however, and the typical visitor would likely notice the walking path and other sites on the map before honing in on the small brown dots. While it is possible to discern from the map, upon close reading, the causal relationship between the walking path and the brown dots situated along it, the inference is ancillary to the more prominent correlation between the walking path and the named sites. While visitors will nonetheless happen upon the sculptures when entering the green spaces, these art installations are clearly secondary to the larger museums and other historic structures depicted in the map. While this may be due to issues of scale, and to the fact that the sculptures physically take up less space than the larger built structures, the map's lack of naming results in what Blair might describe as a silencing: "Texts may also serve to silence or limit other texts by means of their own exclusions" (45). Interestingly, many of the park rangers have come up with their own shorthand names for some of these sculptures, referring to sculptures such as the *Fourteen Hour Clock* (to be discussed later in this chapter) simply as the "clock," perhaps not only recovering them from the map's subtle exclusion but also conveying an implicit closeness with the landscape and its features.

The map is also aesthetically pleasing and paints a tidy picture of the park's boundaries, making the space appear easy to navigate. The map itself is quite well organized and makes clear use of graphical features such as lines, symbols, color, and text. However, when understood within the context of the original mills that stood on this site, the map's tidy representation may be viewed as perpetuating the sort of idealized coexistence of technology and nature that Stanton describes (48). In this sense, the map may be read as an unwitting heterotopia of compensation, perhaps fostering a sort of concessionary space "as perfect, as meticulous, as well arranged as ours is messy, ill constructed, and jumbled" (Foucault, "Of Other Spaces" 27). That is, the map's graphical features aid in the representation of the park as an aesthetically pleasing, cohesively organized arrangement of spaces. But as many of the Mill Girls' narratives reveal, such depictions of mill life were tidy only on the surface, and many writings often revealed subtle descriptions of the difficult work environments that they endured. For example, a letter written by a mill operative named Susan portrays some of the strongest descriptions of how noise pollution at the mills affected the bodies of the Mill Girls. While she clearly attempts to downplay its impact, the loud work environment clearly affects her physical well-being:

> At first the hours seemed very long, but I was so interested in learning that I endured it very well; and when I went out at night the sound of the mill was in my ears, as of crickets, frogs, and jewsharps, all mingled together in a strange discord. After that it seemed as though cotton-wool was in my ears, but now I do not mind at all. You know that people learn to sleep with the thunder of Niagara in their ears, and a cotton mill is no worse, though you wonder that we do not have to hold our breath in such a noise. (Eisler 52)

Here, Susan not only speaks of the incompatible sounds of nature and industry "mingling" together in her head but also conveys that she has adapted to the ringing in her ears. Her letter seems to serve as "thinly disguised reportage," and her narrative seems to serve the more subversive purpose of resisting mill owners' attempts to paint an idyllic portrait of mill life (Eisler 43). Publications in the *Lowell Offering*, such as this one, reflect the tension between the Mill Girls' desire to tell their stories and the possibility of the mill owners' attempts to control those stories. As a result, the *Lowell Offering* seems to function as a disciplinary mechanistic extension of the mill owners' control of both the landscape of the mills and the bodies of the Mill Girls. This disciplinary mechanism then takes the form of a heterotopia of compensation that may also be read as reflected in the park map's visual depiction of the seamless, measured, and even aesthetically pleasing interactions between technology, industry, and the natural environment. These relationships may be inferred through the map's use of graphical features and the interactions of symbolic colors like brown, blue, and green, and the many natural spaces and sites commemorating industrial histories to which they refer.

ARRIVING AT AND ENTERING THE PARK

Upon arriving at the park, visitors will be taken aback by the first set of main arches that lead toward the visitor center (Figure 3).

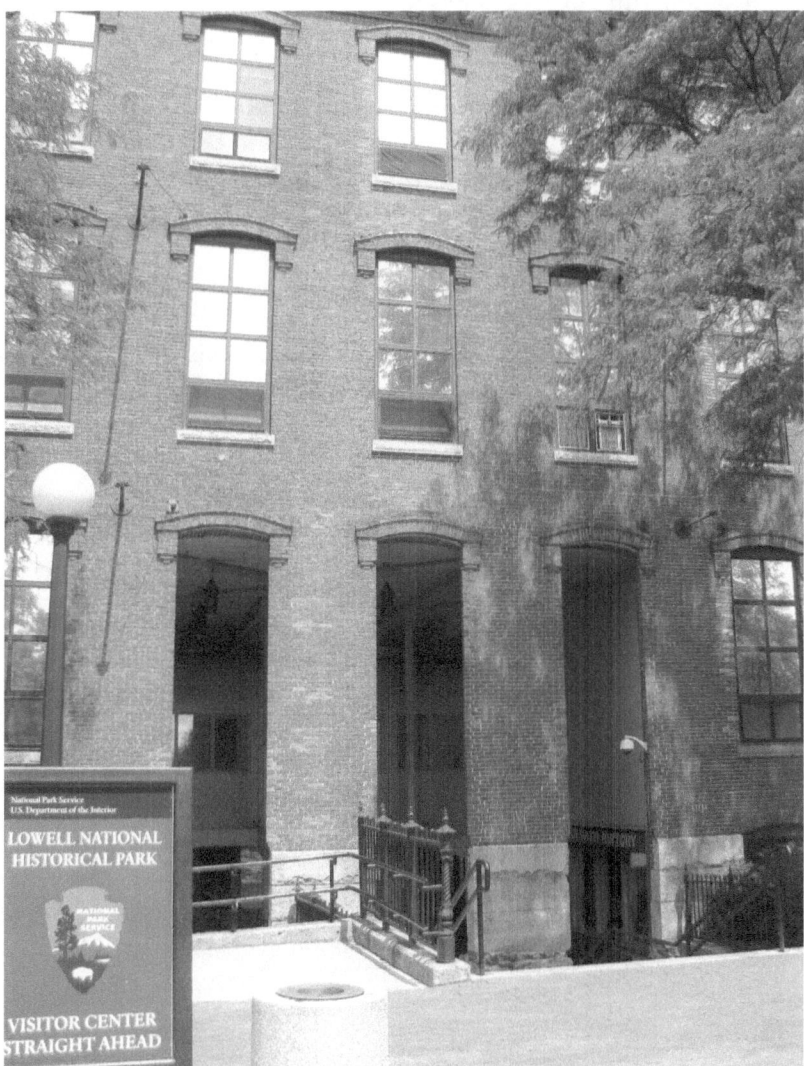

Figure 3: Visitor Center: Main arches, approaching visitor center from parking lot. LMNHP. (Photo by the author.)

The park ranger I interviewed noted that many visitors are indeed "struck by the size of the buildings," and that they note the "the high ceilings [. . .] the brick [. . .] the physicality of the buildings." This feeling of being overwhelmed by the brick buildings is also consistent with many of the Mill Girls' accounts upon arriving at the mills. In a letter written home, for example, the Mill Girl named Susan describes her initial impression upon arriving at the mill: "[I]t all appears very romantic to me. The driver carried me to the 'corporation,' as it is called; and which, so far as I now can describe it, is a number of short parallel streets with high brick blocks on either side" (Eisler 46). Another Mill Girl also describes the mill's "huge sides of brick and mortar" (77). Additionally, the ranger noted that the Mill Girls likely would have been even more overwhelmed than the average park visitor by the high brick buildings, because prior to arriving in Lowell, most Mill Girls had not seen a building much taller than a small town church.

The visitor center is just under and past the main arches (Figures 4–6), inside Market Mills (formerly the Lowell Mills).

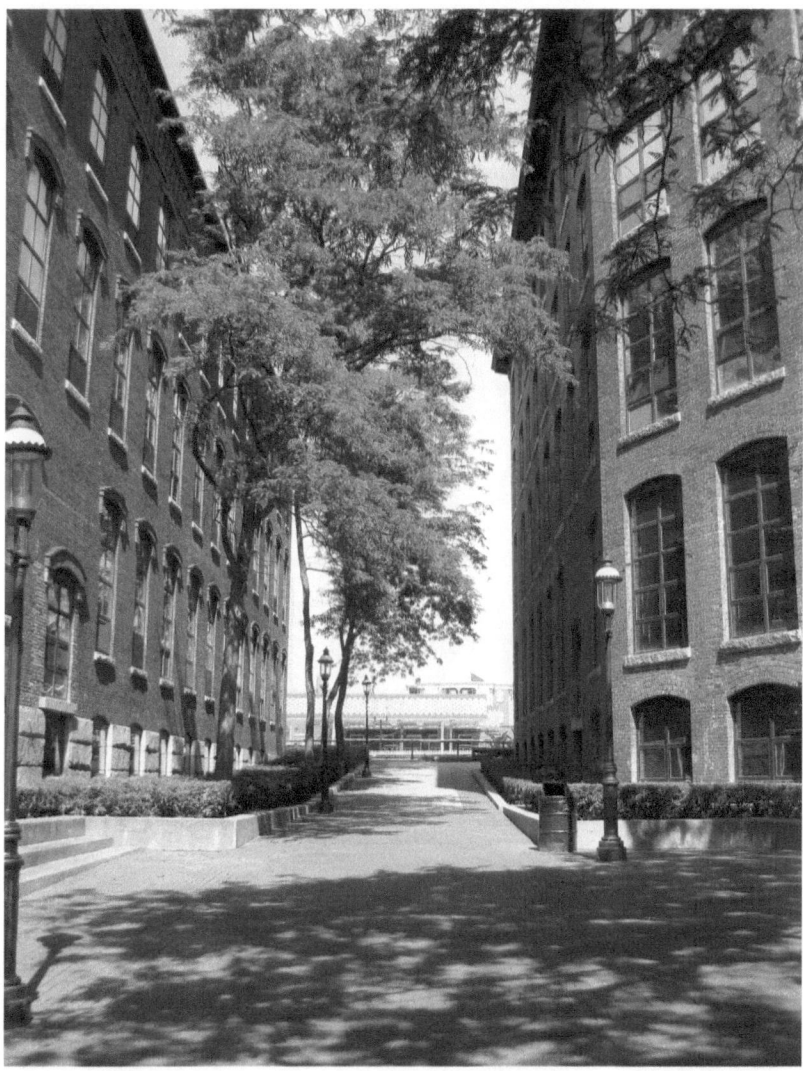

Figure 4: Visitor Center: Just under and past the main arches, looking west. LMNHP. (Photo by the author.)

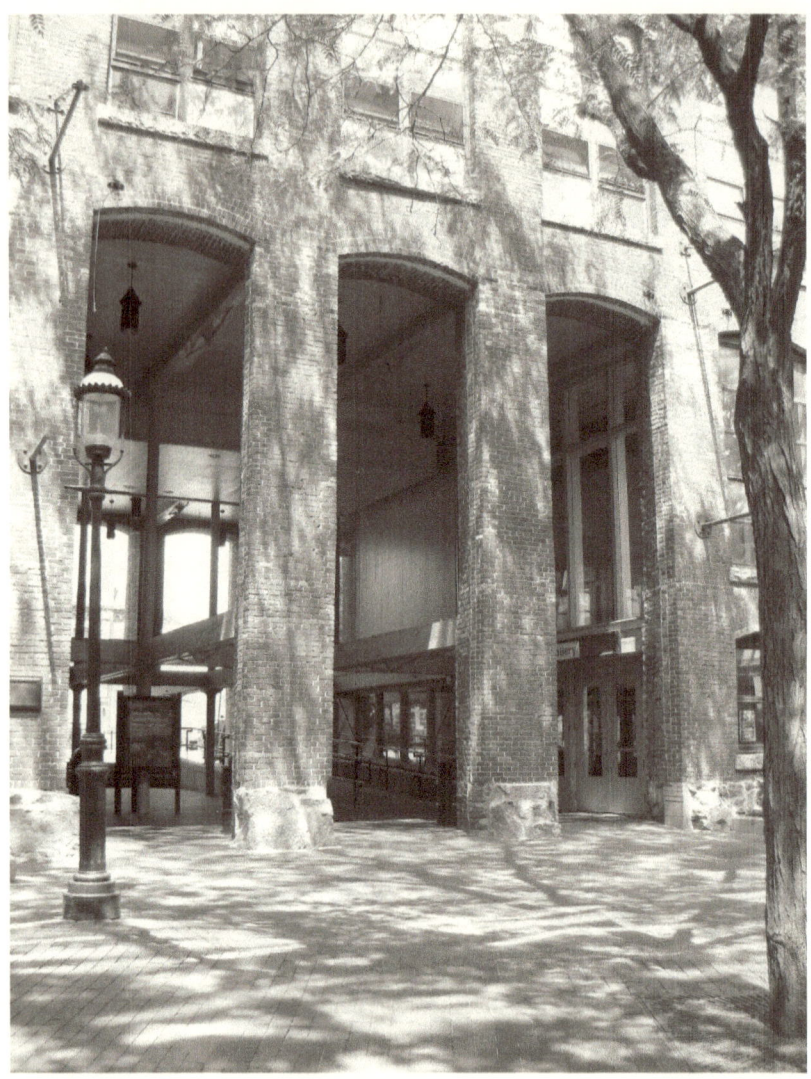

Figure 5: Market Mills: Approaching the visitor center. LMNHP. (Photo by the author.)

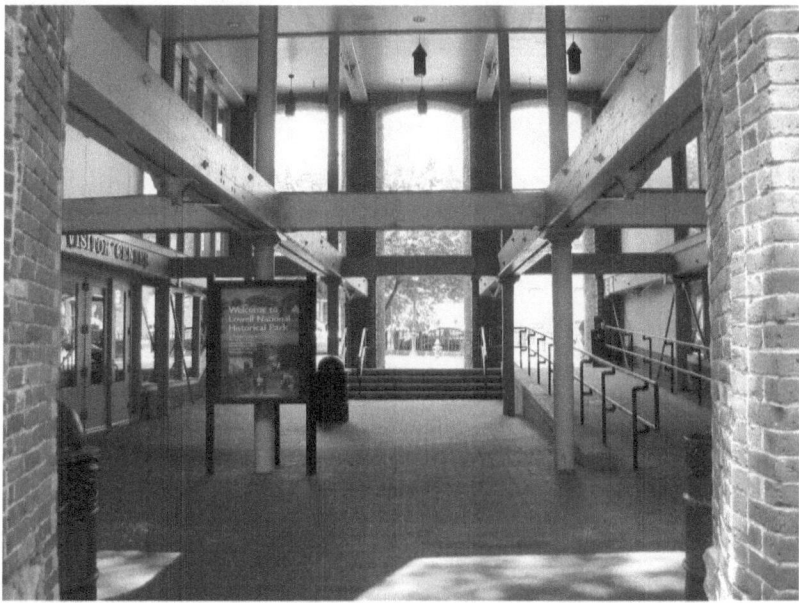

Figure 6: Market Mills: Under the arches with visitor center to the left. LMNHP. (Photo by the author.)

Entering the visitor center from the main parking lot on the southwest side, visitors receive the official park map and any additional information they may require. Leaving the visitor center, they continue under the arches, exiting on the northeast side (on Market Street) to begin the walking tour (Figures 6 and 7).

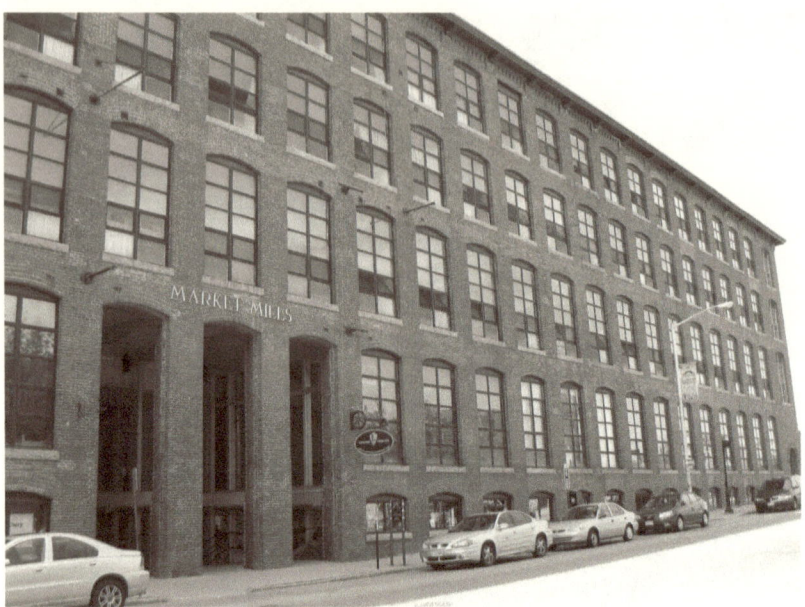

Figure 7: Market Mills and Visitor Center: View from the northeast, Market Street side. LMNHP. (Photo by the author.)

The walking path implicitly relies on a Western schema for reading, and likewise implicitly proceeds northeast and essentially clockwise from the visitor center. Visitors who walk in this direction will first come across Lucy Larcom Park, then Boardinghouse Park, and then eventually the *Homage to Women* memorial. After exiting the visitor center, the first of many wayfinding signs guides visitors across Market Street, where they will find the "beginning" of the Canalway walking path.

Navigating the Park and Its Walking Paths

Along the Canalway walking tour, visitors will come across several wayfinding signs situated along the path. Wayfinding may be defined as "the purposeful, directed, and motivated means for travelling from a point of origin to a given destination. It involves selecting and following set pathways through an existing network," and is thus composed of both "movement" and "decision-making" (Xia et al. 446). Wayfinding signs, or signposts, serve in many contexts to help guide tourists, local citizens, and other visitors from one point to the next. At the LMNHP, wayfinding signs guide visitors along the Canalway walking tour. These wayfinding signs serve as small maps and make use of semiotic concepts to help guide visitors along the walking tour.

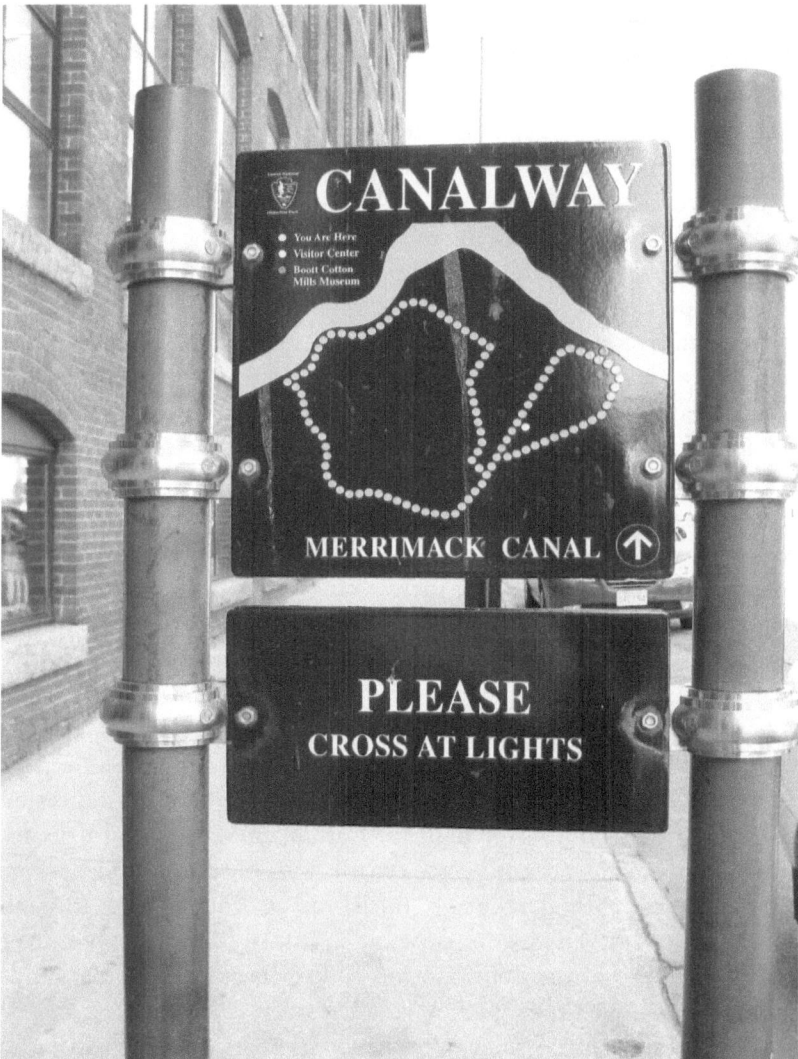

Figure 8: Wayfinding Sign. LMNHP. (Photo by the author.)

The signs iconically depict the Merrimack River and the Lowell canal system by bearing direct similarity to the actual shape of the physical terrain of the river and canal system. The river is depicted with a thick, blue, wavy line along the top of each sign, while the canal system is depicted with a blue dotted ring, symbolizing both water and its relationship to the walking path. Both the representation of the river and the canal system also function symbolically through the use of the color blue to correlate with the "natural" characteristics of water. The blue dots trace the Pawtucket, Northern, and Western Canals, as well as the Merrimack and Eastern Canals. The representation of the full shape of the canal system with the Merrimack River to the north helps spatially situate the visitor more holistically, thus allowing them to form their own cognitive map of the park. Cognitive maps are "representations formed within the mind," and help in the development of spatial understanding and spatial ability; they are often considered "an essential part of wayfinding" (Xia et al. 446). As Xia et al. also note, "The cognitive map can be built through personal experience of the physical world or through reading and learning route maps" (447). The Canalway signs function collectively as a pathway that helps to guide visitors through the park and create their own personal experiences or cognitive maps of the area.

While the Canalway walking path is implicitly included in the dotted loop on the right-hand side of the sign (it is possible to compare the shape of the loop with that of the walking path on the map), it is not explicitly named as such; rather, visitors must read "Canalway" on the sign and subsequently discern that this indeed marks the same walking path designated by the bold brown line on the map. This is not a difficult task, but does require either some comparison between the sign and the park map, or perhaps prior guidance from park rangers. A white dot also appears as a marker along the dotted line, indicating the visitor's position on the path in relation to their general distance (there is no scale) from the visitor center. A yellow dot indicates where they are on the path, described in the legend as "You are here" (though some of these dots are missing on some signs), and a red dot notes their position relative to the Boott Cotton Mills Museum, a popular destination for many visitors.

While the signs do situate the visitor in relation to their distance from the visitor center, they do not provide any additional, specific information about the location of specific features or green spaces within the park. If a park visitor were only interested in walking leisurely along the path or in finding the Boott Cotton Mills Museum, then the wayfinding signs would likely provide enough context and guidance to do so. However, the park ranger I interviewed did note that many visitors come to the park "with a fair amount of pre-knowledge," and have interests in labor history or American history, or have family members who worked at the Lowell Mills. Such visitors want to put themselves in the mindset of their ancestors, as the ranger described it, and may then be interested in more details than the wayfinding signs provide. That said, the wayfinding signs seem to guide the visitor in a general sense; visitors must then set them within the context of other park features to glean more specific information about the park and its artifacts. In this sense, the wayfinding signs function as part of the larger intertext of the park.

The lack of specificity in the wayfinding signs may also encourages the open-ended manner in which the park may be navigated. That is, many visitors find that the park, on the whole, does not seem to have a clear beginning or a clear end. Visitors may focus on one aspect of the park or walk the entire path, making the map even more powerful in its ability to point out important features and direct visitors' movements along the way. I asked the park ranger if she or other guides focus on specific areas of the park while giving tours. She responded that tours given by park rangers tend to vary based on the specialty area and interests of the ranger. Consistent with the open-ended layout of the park, the ranger noted that visitors can really walk in whichever direction they choose, and while park rangers are there "to facilitate connections," it is really up to the "visitors [to] make their own connections." While the various exhibits and green spaces are certainly conducive for making the sort of connections to which the ranger refers, they can pose a challenge for visitors as well. Thus, the park ranger noted that many visitors critique the park's layout, pointing out that they must visit all the exhibits to get the full picture. Because the exhibits are very compartmentalized and focus on specific topics, she says, "folks don't feel like they're getting the big picture" unless they make the time to "see everything." This is certainly true enough; if visitors come to the park without a specific agenda, its open-ended layout can easily make for an overwhelming experience, or at least make it difficult to discern a holistic picture of the site.

Blair notes that some memorial sites will "suggest—sometimes *prescribe*—*pathways* for a visitor to traverse, and those pathways influence reception significantly" (47). Interestingly, the significance of this park's layout seems to be its lack of an explicitly imposed pathway. Certainly, the walking tour prescribes a pathway to an extent, but its clearly open-ended nature and how it weaves through and around the downtown vicinity both afford it myriad navigational choices at best, and make it perhaps overwhelming at worst. Indeed, aspects of the park can have the consequence of feeling overwhelming to visitors. And as mentioned earlier, the Mill Girls too described feeling overwhelmed upon initially entering the mills and boardinghouses. For example, of her first meal in the boardinghouse, the mill worker named Susan writes: "You can hardly think how my heart beat when I heard the bells ring for the girls to come to supper, and then the doors began to slam, and then Mrs. C. took me into the dining-room, where there were three common-sized tables, and she seated me at one of them, and then the girls thickened around me, until I was almost dizzy" (Eisler 47). From this narrative, it is clear that the sheer density of bodies is anxiety-producing or almost "dizzying," as this writer puts it. Interesting also is the way in which the anaphoric series and polysendetic repetition of "and then" slows down the flow of the passage, giving it a weight that not only conveys Susan's anxiety but also mimics the overwhelming environment of the dining room. Perhaps, then, the potentially overwhelming or open-ended nature of the park's layout does not do its visitors any favors, and why should it? If the influence exerted by the physical environment of the mills and boardinghouses indeed acted on the minds and bodies of the Mill Girls in such ways, then perhaps part of the park's rhetorical impact resides in its ability to invoke in its visitors feelings similar to that which the Mill Girls might have experienced.

The park is built right into the landscape of downtown Lowell and comprises an area of only about three square miles. As Cathy Stanton notes in her study of Lowell's public history, the park is atypical in terms of other national parks, in that

> it is not a neatly bounded piece of real estate owned outright by the National Park Service, but a series of buildings and open spaces within the downtown area and along the canal system that once powered the textile mills. [. . .] Visitors—and even local residents—are often still confused about where the park actually *is*. And so most park rangers begin their guided tours by addressing that point. [. . .] [W]hat is being interpreted at the national park is not just isolated mills or canals or corporation boardinghouses, but an entire city, whose life today is in many senses an outgrowth of its industrial past. (3)

The park indeed retains the characteristics of its industrial past. To walk along the low, narrow sidewalks is to walk almost flush with the cobblestone, brick streets that pave the city center. Most of the brick buildings of the downtown area are quite narrow and compressed, and still retain the visual characteristics of Lowell's industrial heritage. On the one hand, visitors beginning the self-guided walking tour may, as Stanton notes, wonder where or what the park actually is, making it difficult to decide where to go next. On the other hand, the system of wayfinding signs and informational plaques, as well as the many local citizens, park employees, and visitors make it difficult to get "lost" in the sense of being unable to discern one's way.

One consequence of the park's layout and its immersion in the city center is that visitors get the feeling that they are never truly alone. Likewise, the Mill Girls were never alone—even when they worked the looms, lost in their thoughts, or distracted by the loud noises of the machinery, they were always in the company of other workers and being overseen by chaperones. As Foucault writes, this notion of the managed body was part of a new scale of control in the eighteenth century "projects of docility" that exercised "a subtle coercion" at the level of "movements, gestures, attitudes, rapidity: an infinitesimal power over the active body" (Foucault, *Discipline* 137). In this sense, the park, or really the entire downtown Lowell historic district, still exercises a subtle control over the active body. And as visitors navigate the park, meant to depict and describe eighteenth century industrialized Lowell, they likely feel anything but alone. They may feel at once overwhelmed by the navigational possibilities, and yet controlled by the narrow, cobblestone streets and the tall brick buildings that obstruct the view of the horizon. The entire district feels somber, almost haunted.

LUCY LARCOM PARK

As visitors cross Market Street, following the Canalway walking path to the Merrimack Canal, they will enter Lucy Larcom Park. A wayfinding sign also marks the south entrance of the park (Figure 9).

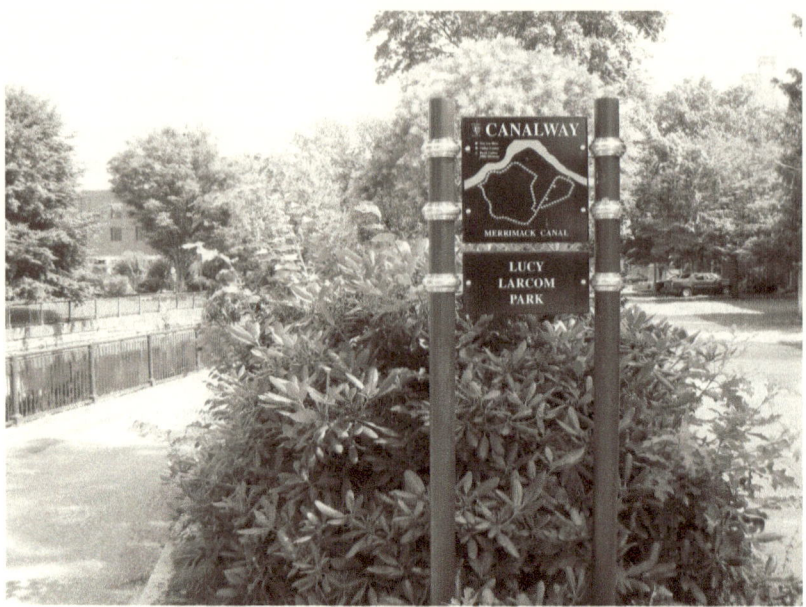

Figure 9: Wayfinding Sign: Lucy Larcom Park. LMNHP. (Photo by the author.)

Lucy Larcom Park is a clearly-delineated, linear, rectangular stretch of green space. Approximately 1/5 mile-long, it runs alongside the Merrimack Canal (Figure 10).

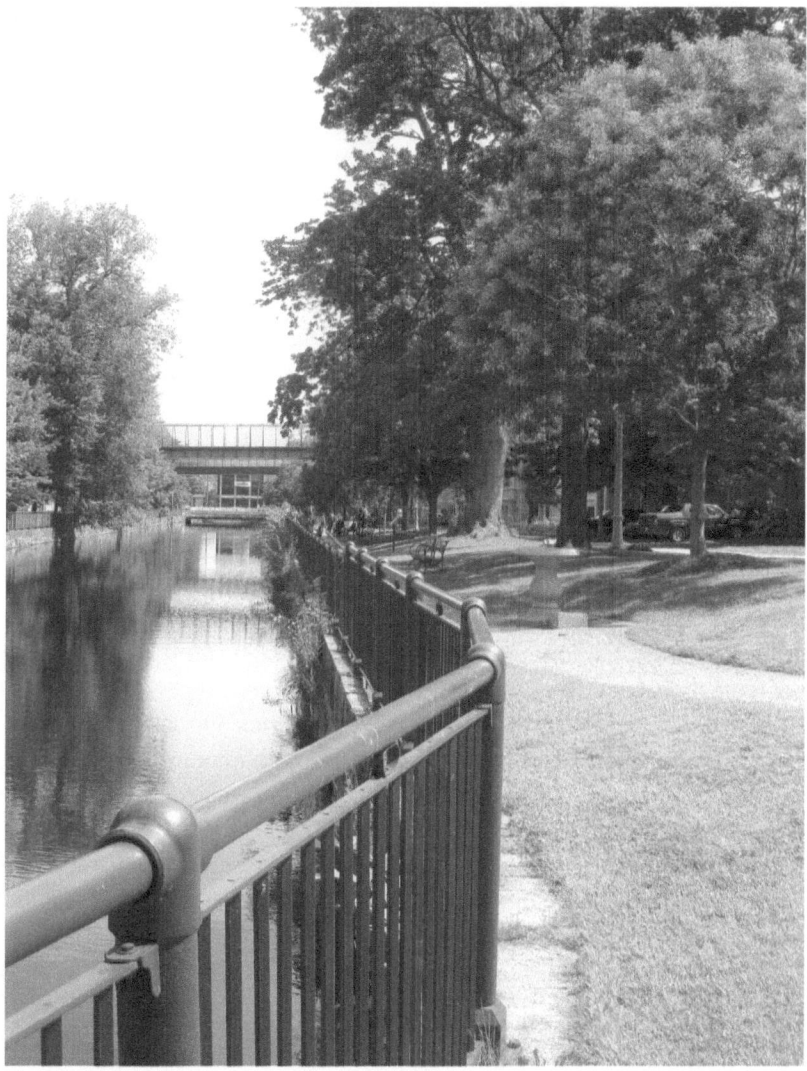

Figure 10: Lucy Larcom Park: View of park facing northeast. LMNHP. (Photo by the author.)

The park ostensibly commemorates the later Mill Girl (labor reform) era, but may also be understood as commemorating the women who worked in the mills during the early to mid-1800s. The park contains an art installation comprised of four main elements or sculptures. These sculptures focus primarily on the political work of former Mill Girl-turned-labor activist, Sarah Bagley, and the mill worker, Lucy

Larcom, who was famous not only for her poetry but also for being one of the youngest women workers, having come to the mills at the age of eleven. The four sculptures in Lucy Larcom Park commemorate the founding of the Lowell Female Labor Reform Association (LFLRA) and the petition for the Ten-Hour Movement; they are inscribed with quotations from the LFLRA as well as quotations from Sarah Bagley and Lucy Larcom.

In December of 1844, Sarah Bagley along with five other women formed the Lowell Female Labor Reform Association "to improve health conditions and lobby for the Ten-Hour Day" ("Sarah George Bagley"). In 1845, Bagley and her colleagues "petitioned the Massachusetts Legislature, demanding a Ten-Hour Day" ("Sarah George Bagley"). The Ten-Hour Movement lobbied for legislation preventing mill owners from requiring female laborers to work 12–14 hour days, requiring instead no more than a ten-hour day.[4] Under Bagley's leadership, the LFLRA also published the *Voice of Industry,* a labor reform newspaper whose tone and content was much more direct and politically-charged than that of the *Lowell Offering*.[5] Sculptures in Lucy Larcom Park thus contain quotations from the *Voice of Industry* and sections of legislation published during this time.

Lucy Larcom Park was designed by Frederick Olmsted, the nineteenth century American architect who is perhaps most well-known for designing Central Park in New York City. Olmsted advocated design that included native plants and remained true to the natural environment (Park Ranger). The park is home to a public art installation comprised of five main elements built of granite and steel, designed and installed from 1995–1996 by the artist Ellen Rothenberg. The park incorporates native plants and trees, and the juxtaposition of the granite and metal sculptures within this natural and native setting is sometimes perceived as reflecting the Boston Associates' attempts to create a portrait of the "ideal manufacturing community," in which American democratic ideals were reflected through the seeming coexistence of industry and the natural environment (Stanton 48). In coming to work at the mill, the female operatives sacrificed a great deal of their ties with nature, and often spent most of their time in noisy workrooms. While the mill landscapes included flower gardens and while the Mill Girls had some access to rural settings, "the experience simply did not compare to living and working on the women's native homesteads [. . .] and it played an important part in the women's growing disenchantment with mill labor" (Montrie 21). Lucy Larcom Park in many ways performs the Mill Girls' growing disenchantment with mill life.

To walk through the park is to experience a series of granite and steel sculptures immersed within a natural environment that is at once peaceful and flowing, and linear and constructed. To stand at one end of the park and take in its visual linearity punctuated with the atonality of its material sculptures makes for a dissonant experience that is at once serene and destabilizing. The sculptures in the park, which together comprise a cohesive public art installation, also represent the increasingly harsh work conditions endured by the Mill Girls, and the subsequent protests that characterized their fight for labor reform. Through the park's performance of the merging of industry and nature, visitors begin to experience the crisis heterotopias of the Lowell Mills.

Consistent with the park's open-ended layout is the fact that the five sculptures in Lucy Larcom Park, arranged lengthwise along the rectangular green space, are designed to be experienced walking from either direction. The park's imposed linearity and parallelism make for what Blair refers to as a "prescribed pathway," though one that may be approached from either direction (47). Lucy Larcom Park also has a more calming influence, both in its distance from the taller brick buildings that are so prevalent in other areas of the park, and in its running alongside this quieter section of the Merrimack Canal.

Entering Lucy Larcom Park from the south side and walking northeast, visitors will initially come across an informational or interpretive plaque that introduces the public art installation housed there (Figure 11).[6] Comprised of five sculptures, the installation is called *Industry Not Servitude.*

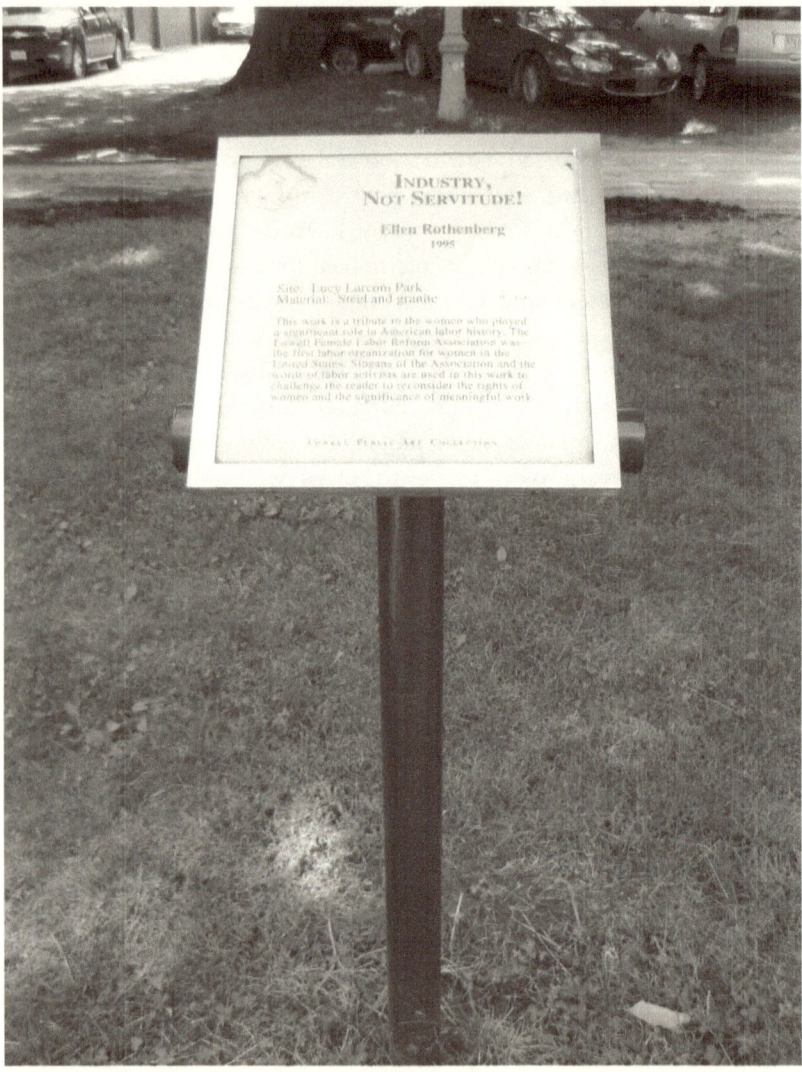

Figure 11: Informational Plaque: *Industry Not Servitude*. LMNHP. (Photo by the author.)

The plaque describes the theme of the five elements as follows:

> A tribute to the women who played a significant role in American labor history. The Lowell Female Labor Reform Association was the first labor organization for women in the United States. Slogans of the Association and words of labor activists are used in this work to challenge the reader to reconsider the

rights of women and the significance of meaningful work. (Lowell Public Art, "Industry Not Servitude")

Heading northeast through Lucy Larcom Park, visitors will then encounter the first installation of *Industry Not Servitude,* a sculpture titled *Steps.*

Steps

The *Steps* sculpture (Figure 12) is comprised of two granite stairs built into the sloped landscape of the park. The steps are engraved with a quotation from the LFLRA: "Try again! Industry, not servitude. As is woman, so is the race" (Rothenberg, *Steps*). This sculpture's material existence seems to enact a largely symbolic function that also reflects a message pervasive throughout the green space and the other pieces in the installation; that is, while it is certainly feasible that visitors may engage with the sculpture by attempting to climb these two steps, it seems more likely that the sculpture, integrated smoothly with the park's landscape, invites its viewer to observe it and discern from its textual and physical message the idea of perseverance and the relentless pursuit of labor reform.

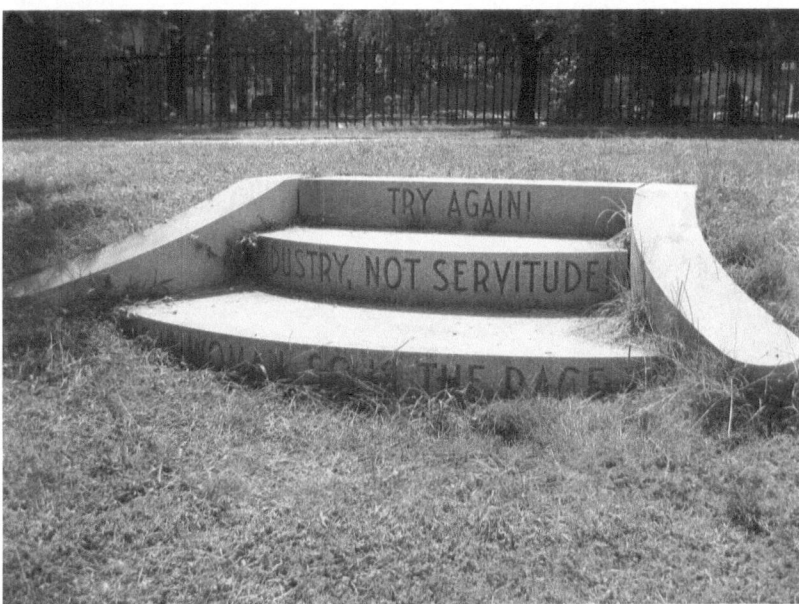

Figure 12: *Steps* Sculpture. LMNHP. (Photo by the author.)

Because the steps are fairly shallow and because the sculpture is so embedded in the landscape, it is easy to miss and figures less prominently on the landscape than some of the larger or more spatially obtrusive sculptures in the installation. Comprised of two full steps, as opposed to, say, three or four, it appears more subtle and sits low to the ground. In the winter months, it may easily become covered with snow and thus not be visible at all.[7] Nonetheless, the sculpture conveys the idea of ascendance and even transcendence of current circumstances, consistent with the message performed by the installation more generally. Moreover, the physical challenge of climbing stairs is described by several of the Mill Girls in their narratives of mill and boardinghouse life. In one example, the Mill Girl named Susan describes her impressions when first being shown to her room in the boardinghouse: "I was shown up three flights of stairs, into what is called 'the long attic'— where they put all poor stranger girls—the most objectionable places being always left for new comers. There were three beds in it, only two of which were occupied, for this is always the room for vacancies" (Eisler 48). Implicit perhaps in Susan's narrative is the physical work of ascending three flights of stairs only to discover, to her dismay, that she is to room with three other Mill Girls in attic quarters. Thus, perhaps it is doubly fitting that *Industry Not Servitude* incorporates this gesture to the physical hardships of the Mill Girls, and accomplishes through its message of perseverance a subtext that describes the challenges they endured on a regular basis.

Just past the steps is the second sculpture, officially titled *Circular Fence Sculpture with Poem*, though informally referred to by the park ranger and other park guides as the "Cage Around the Tree," a name that also describes the piece well.

Circular Fence Sculpture with Poem

The *Circular Fence Sculpture with Poem,* also part of *Industry Not Servitude* (see Figure 13), is composed of a circular, steel, cage-like structure that encloses a single tree. The steel cage is inscribed with the words of a poem written by Lucy Larcom; however, the exact name of the poem remains unclear.[8]

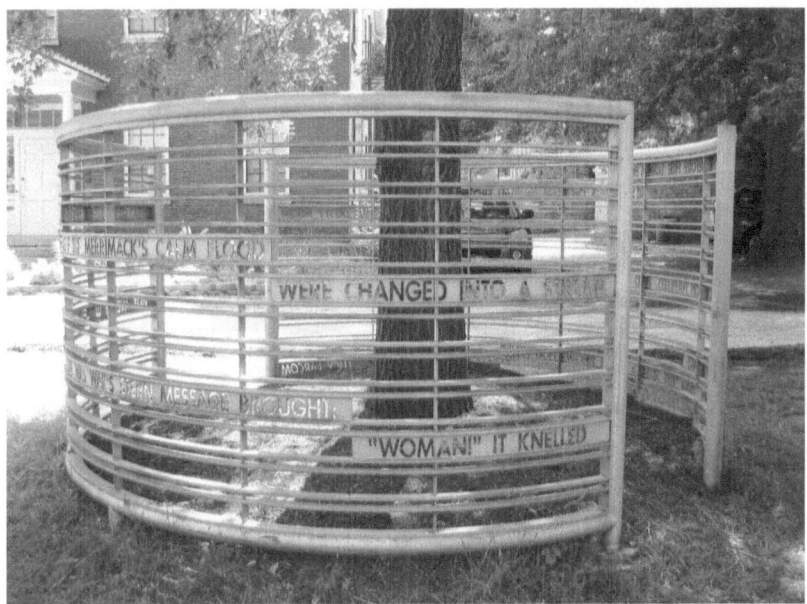

Figure 13: *Circular Fence Sculpture with Poem.* LMNHP. (Photo by the author.)

In reading the poem or in viewing the sculpture, the reader is unsure where to begin, or from which direction to read. The non-linearity of the text is disorienting when set against the park's linearity, and makes for a viewing experience that invokes the tension felt by the Mill Girls as they worked in a chaotic, crisis heterotopia that was presented to the outside world as a neat and tidy heterotopia of compensation. While I transcribed the poem word for word as best I could, I feel it bears noting that my own role here was not of the typical observer, as most observers would likely approach the sculpture but not transcribe its contents. My transcription then reads as follows:

> So up and down before her loom
> She paces on, and to and fro
> Til sunset fills the dusty room of blood
> Too soon fulfilled and all too true
> The words she murmured as she wrought:
> From heart to heart
> And makes the water really glow
> As if the Merrimack's calm flood

> Were changed into a stream
> But weary, weaver, not to you
> Alone was war's stern message brought:
> "Woman" it knelled
> (Rothenberg, *Circular Fence*)

The sculpture itself was designed by Rothenberg, but the poem, transcribed above, was presumably written by either Sarah Bagley or Lucy Larcom. The poem describes the oppressive and dangerous working conditions of the mill operative who works at her loom for extended periods of time. Interesting too is the reference to the Merrimack's "calm flood" changing into a "stream," as many narratives from the *Lowell Offering* describe the mill worker who longs to sit by the glowing or glistening stream on the farm on which she grew up. For example, in this melancholic passage from the narrative, "A Week in the Mill," a Mill Girl describes her longing for the natural landscape of her rural home: "Every thing having been cleaned and neatly arranged on the Saturday night, she has less to occupy her on Monday than on other days; and you may see her leaning from the window to watch the glitter of the sunrise on the water, or looking away at the distant forests and fields, while memory wanders to her beloved country home" (Eisler 75).

In addition to the poem's powerful message and disorienting non-linearity, the sculpture into which it is carved is quite disconcerting. Even if visitors do not read the contents of the sculpture in its entirety, which, based on my observations, many do not, they likely still feel the tension of the cage and notice the writing that encompasses it. To walk around the structure while reading is also disorienting and dizzying. The experience is especially taxing on the eyes, in terms of the general sense of disequilibrium that comes from reading with one's head tilted sideways while walking back and forth. An opening also exists in the cage that allows visitors to experience the piece from the inside. To stand inside the cage not only makes it nearly impossible to read the text of the sculpture but also represents the Mill Girls' interaction with their physical environment and the ways in which they came to feel trapped within an industrial setting that left them longing for the natural settings to which they were more accustomed. In this way, to experience the sculpture is to attend to the material manifestation of its symbolism (Blair 19). Visitors of this sculpture

find themselves trapped within the irony of a physical, metal structure immersed within a natural setting; further, to try and read the
text of the sculpture is to engage in a physically taxing process that
exacerbates an already overwhelming experience. By understanding
rhetoric as not solely symbolic but also "in part by its potential for
consequence" on the body, it is possible to understand visitors' interactions with this sculpture as a rhetorical process that allows for a more
empathetic understanding of the oppressive conditions in which the
Mill Girls worked (Blair 19).

Significant too is the fact that this sculpture literally encircles a tree.
This enclosure may be read as embodying the Mill Girls' constructed
experience of nature that was imported into the mill landscape: "From
the beginning, mill owners had made some effort to beautify Lowell
as part of their paternalist scheme. They instructed managers to maintain flower gardens outside the factories and, because the mill acreage
was mostly cleared farmland, they oversaw tree planting" (Montrie
24). The cage around the tree performs the Mill Girls' relationship
with nature—constructed, artificial, and limited. Thus while the textual context of the fence may be seen as commemorating the labor
reform era, its physical structure and its impact on the body may be
understood as representing the early Mill Girl period, during which
rural Massachusetts farm girls came to Lowell in search of a better
way of life, resulting in the sacrifice of their sacred relationship with
nature. It is then the interplay of the sculpture's symbolic aspects with
the visitor's personalized and embodied experience of its material impact that affords the artifact its unique and contextualized possibilities
for consequence and more empathetic understanding.

The Path Markers, or "Speed Bump" Sculptures

Continuing on through Lucy Larcom Park, visitors will happen across
the third element of Rothenberg's *Industry Not Servitude* installation,
the first in a series of four separate but related smaller granite *path
markers,* or small sculptures that punctuate the landscape and separate the larger sculptures. These path markers each literally resemble
speed bumps and are often casually referred to as such by park guides
(Figures 14–21).

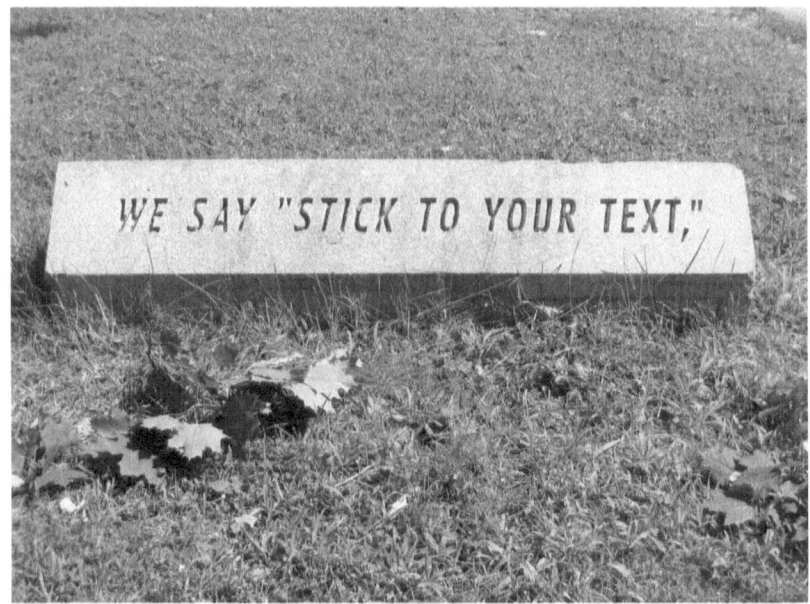

Figure 14: *Path Marker 1*: Front. LMNHP. (Photo by the author.)

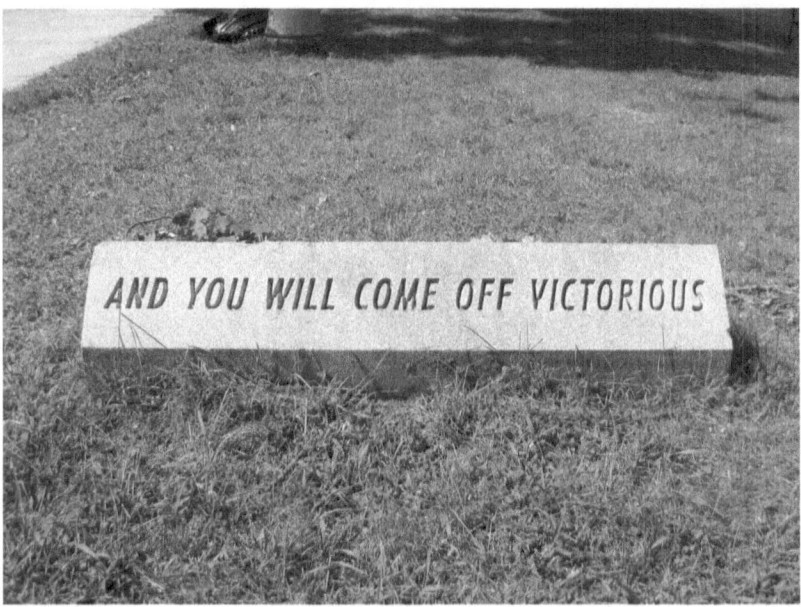

Figure 15: *Path Marker 1*: Back. LMNHP. (Photo by the author.)

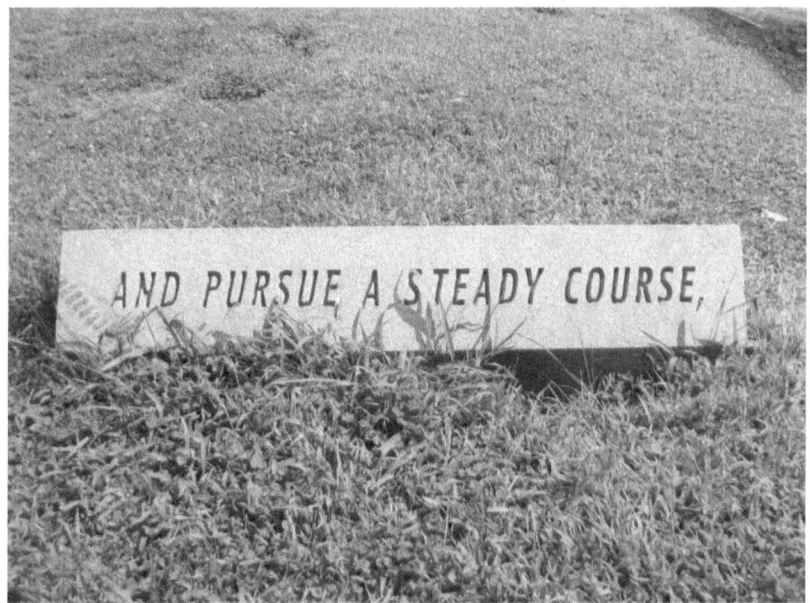

Figure 16: *Path Marker 2*: Front. LMNHP. (Photo by the author.)

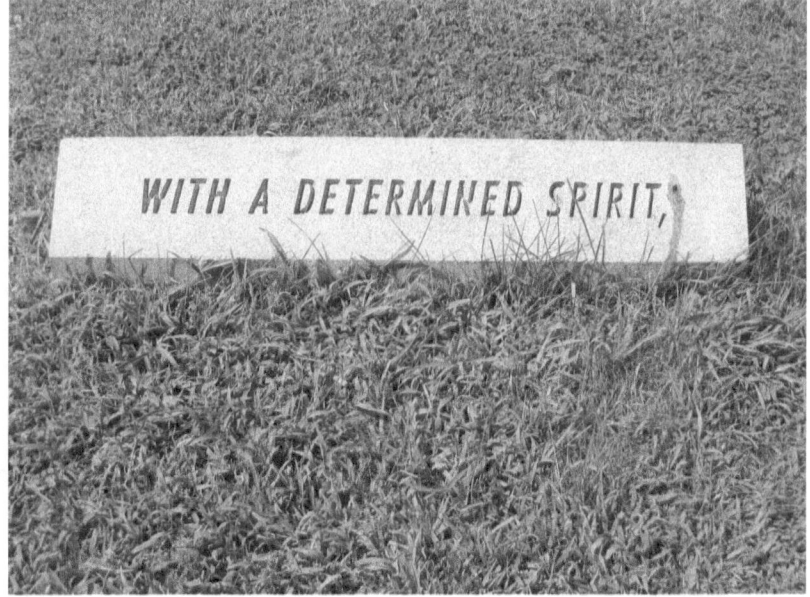

Figure 17: *Path Marker 2*: Back. LMNHP. (Photo by the author.)

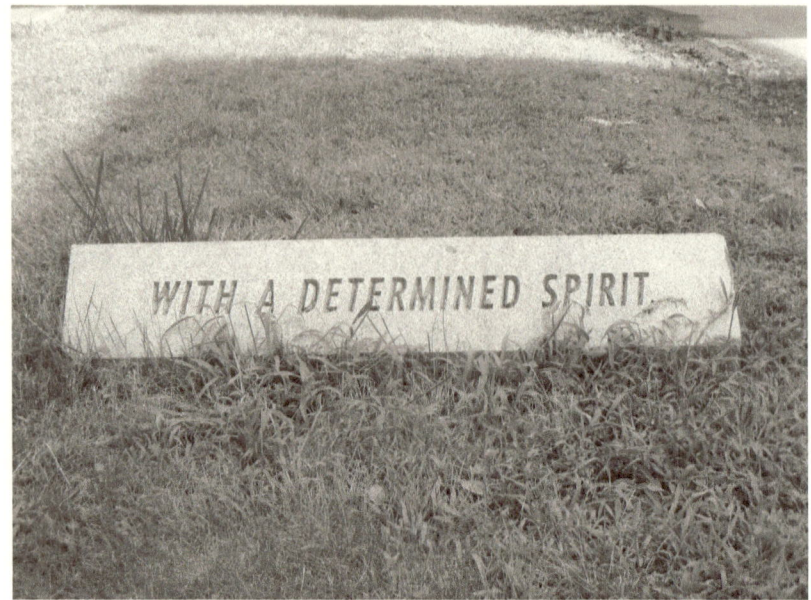

Figure 18: *Path Marker 3*: Front. LMNHP. (Photo by the author.)

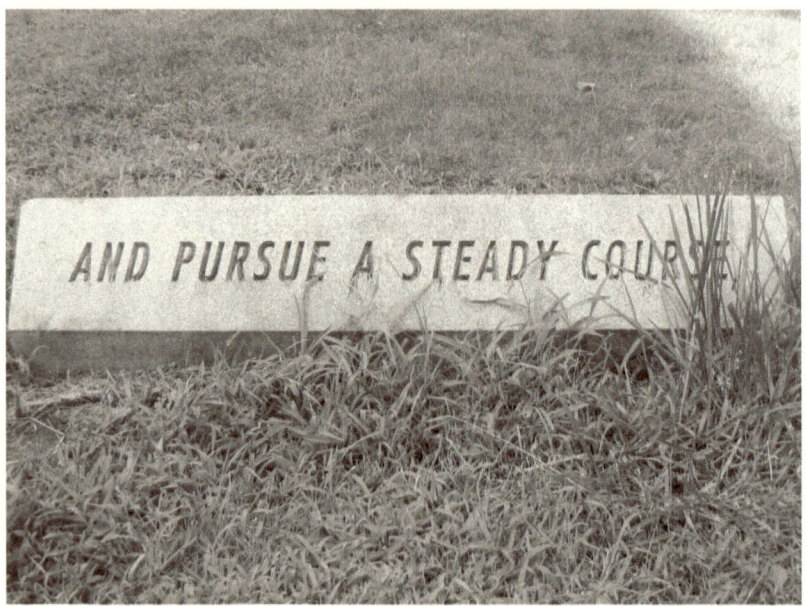

Figure 19: *Path Marker 3*: Back. LMNHP. (Photo by the author.)

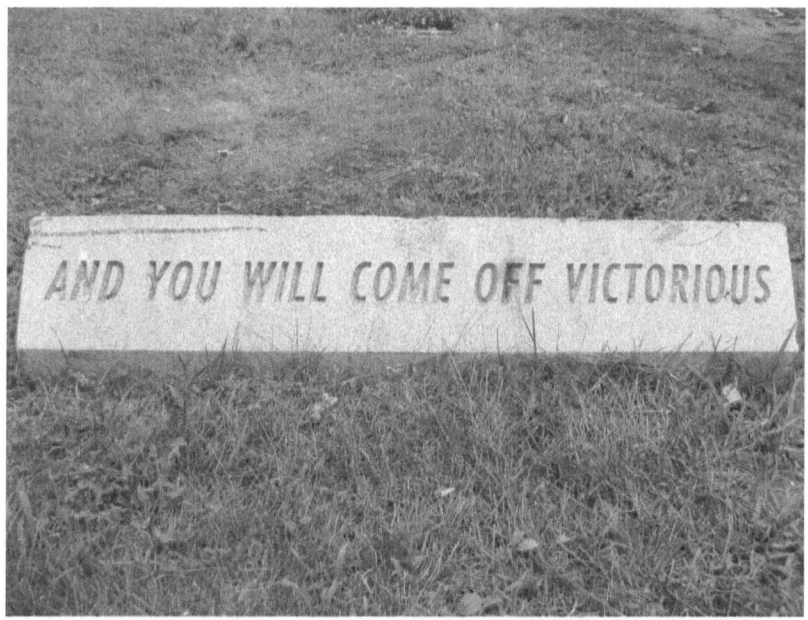

Figure 20: *Path Marker 4*: Front. LMNHP. (Photo by the author.)

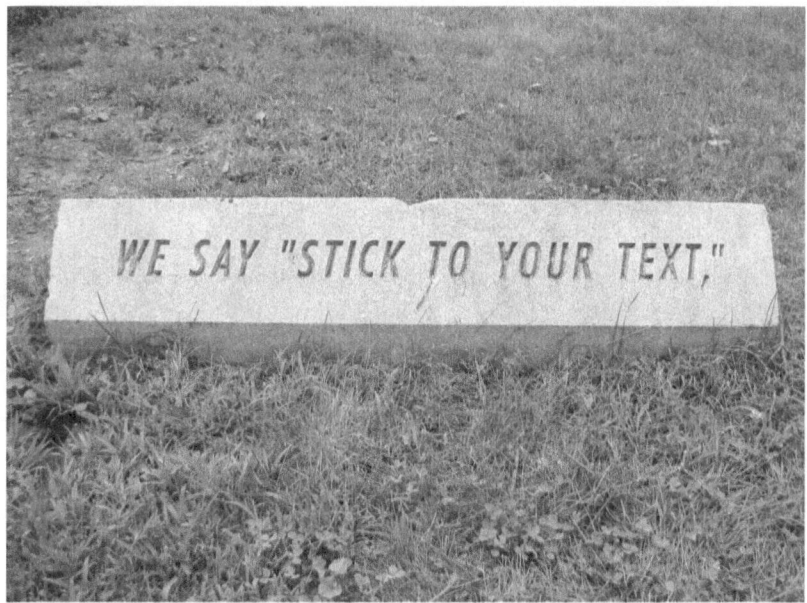

Figure 21: *Path Marker 4*: Back. LMNHP. (Photo by the author.)

The first two path markers appear just after the *Circular Fence.* The third appears after the second larger sculpture, called *Seating Circle: Truth Loses Nothing Upon Investigation,* and the fourth appears after the third larger sculpture, called the *Fourteen Hour Clock.* The elements of this installation thus progress linearly as follows: *Steps; Circular Fence Sculpture with Poem; Path Marker 1; Path Marker 2; Seating Circle; Path Marker 3; Fourteen Hour Clock; Path Marker 4.* The four path markers are two-sided, engraved on each side with quotations that, when read in succession, walking from either direction, read as follows:

> We say "stick to your text,"
> And you will come off victorious.
> And pursue a steady course
> With a determined spirit,
> With a determined spirit,
> And pursue a steady course,
> And you will come off victorious
> We say "stick to your text,"
> (Rothenberg, *Path Markers*)

The sculptures are each engraved with quotations (above) from the writings of Sarah Bagley that presumably first appeared in the *Voice of Industry.*[9] These quotations may be read walking from either direction (or read on paper from beginning to end, or end to beginning). The result is a hybrid figure of repetition, comprised of both parallelism and chiasmus. The reversal of the order of phrases starting in the middle of the poem, such as to end with the same phrase as the beginning, indicates a form of parallel structure, while the inversion of full phrases indicates a variation on the theme of chiasmus (whereas traditional chiasmus would invert only grammatical structures within the sentences or phrases, as opposed to the phrase in its entirety). The parallelism and chiasmus within these sculptures further emphasizes and performs their message. These sculptures perform persistence—be persistent in the course you pursue, they state—persevere in the battle for labor reform. The sculptures, too, do not let visitors out of their way. They look and act like speed bumps, interrupting the visitor as they attempt to move through a seemingly uncomplicated linear landscape. While the park's layout is flat and linear, these sculptures punctuate the landscape and make for anything but a smooth road. The speed bumps stop the visitor; because they are small in size and low to the ground, visitors must take care not to trip over them. Because they're shaped like speed bumps, and because Western culture arguably has a schema for such structures, the tendency is to slow down to read them. In this way, the structures do "obvious work on the body"; they not only "direct the vision" to their textual content, but they also direct and control "the vector, speed, or possibilities of physical movement" (Blair 46). The repetition of the sculpture's text not only helps ensure that the viewer absorbs at least a portion of the message or at least pick ups on the patterns of repetition, but also helps communicate the gravity and emotion of the message.

These path markers may be seen as performing both the crisis heterotopias and the heterotopias of compensation that characterized the early mills. That is, on the one hand, the beauty and linearity of Lucy Larcom Park perhaps emphasizes the mill owners' attempts to present an aesthetically pleasing, sanitized narrative of mill life or the heterotopias of compensation; on the other hand, these sculptures punctuate the landscape and interrupt the visitor's experience by demanding their physical and mental attention, thereby performing the physical difficulties and emotional dissonance experienced by the Mill Girls. Together, the placement and composition of the speed bumps coupled with the larger sculptures, produce layers of destabilizing interruptions in the visitor's physical and mental experience of the landscape. Thus, the speed bump sculptures, as material, visual rhetorics of heterotopic space, help illuminate the physical and mental experiences of the Mill Girls' lives and their battle for labor reform legislation.

The Seating Circle

After encountering the second path marker, visitors will come across the third of the four larger sculptures, called the *Seating Circle: Truth Loses Nothing Upon Investigation* (Figures 22 and 23).

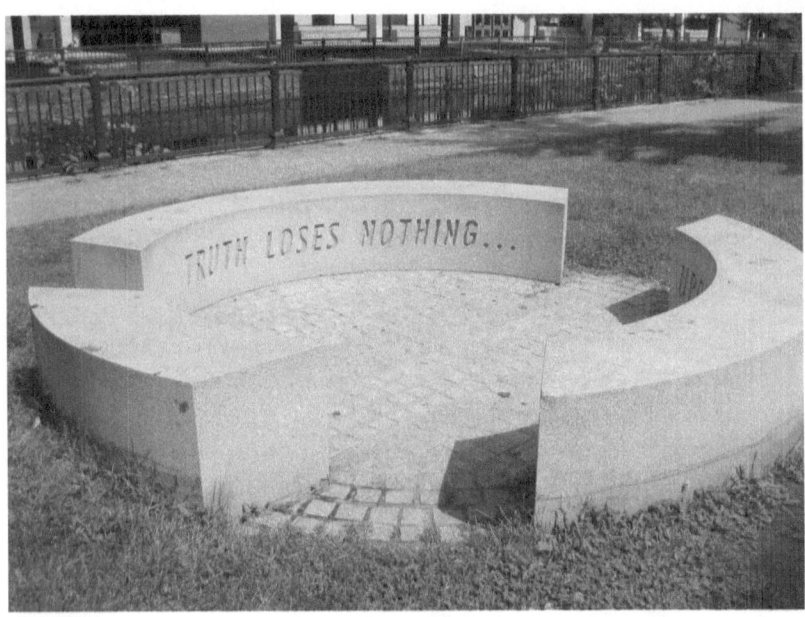

Figure 22: *Seating Circle: Truth Loses Nothing Upon Investigation.* LMNHP. (Photo by the author.)

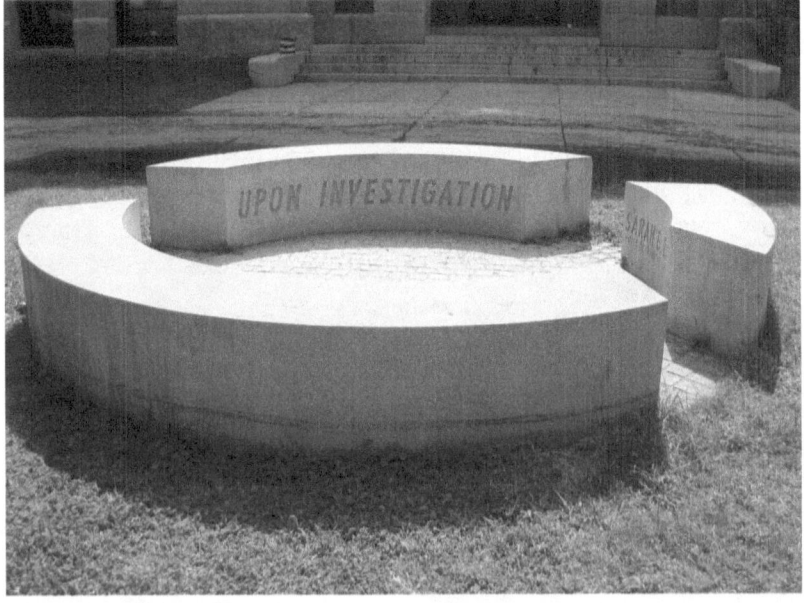

Figure 23: *Seating Circle: Truth Loses Nothing Upon Investigation.* LMNHP. (Photo by the author.)

The *Seating Circle* is comprised of a large granite circle, divided into thirds, but still circular in shape; that is, the pieces of the circle have not been rearranged—just separated. The sculpture sits low to the ground, only about two feet in height, and the inner portion of the circle contains a cobblestone floor with a circumference of about six or seven feet that invites visitors to either enter the circle or treat it as a bench. The largest two thirds of the circle (the three sections are not the same size) are engraved with pieces of a quote from Sarah Bagley, and the smallest third is engraved with her name, thereby attributing the quote to her. The section that faces out toward the walking path, which visitors likely see first, reads: "Truth loses nothing . . ."; the second of the three pieces, moving clockwise around the circle, reads: "upon investigation." The last third is inscribed with the quote's citation: "Sarah G. Bagley, *Voice of Industry,* January 9, 1846" (Rothenberg, *Seating Circle*).

The sculpture's textual content combined with its design and layout make for a message that may again be read as performing the crisis heterotopias endured by the Mill Girls. The broken circle creates a feeling of discontinuity with the visitor. The quotation and its citation, however, span the entire circle, and the textual content provides the sculpture with some continuity. The quotation, "Truth loses nothing upon investigation," comes from the labor reform paper founded and led by Sarah Bagley. The quote suggests that the mill workers have nothing to hide—they do not embellish or exaggerate the conditions of their work environment, and an investigation of these environments will yield information consistent with the truths they tell. This quotation may also be read as understatement. Not only does truth lose nothing upon investigation, but truths are made even more apparent upon investigation, and firsthand experiences of these working environments are likely even more powerful than protesters' verbal accounts. The broken circle indicates the various disconnects between mill workers, superintendents, and mill owners; however, the message carved in granite within each section of the circle suggests that continuity is still possible and may be achieved with participation from governing bodies and upon investigation into the mill environment. Consistent with the message of the speed bumps, then, proponents of labor reform mustn't give up hope. In fact, the two path markers or speed bump sculptures that bookend this sculpture read: "And pursue a steady course [. . .] With a determined spirit," and "With a determined spirit, [. . .] And pursue a steady course."

The *Seating Circle* also invites the viewer into its discontinuity. Like the *Circular Fence,* once inside the circle, it becomes more difficult to read the full text. Visitors must turn 360 degrees as they read, which again makes for a dizzying experience. The cobblestone floor within the circle is nonetheless inviting and aesthetically pleasing, and so the sculpture sends a mixed message that is both encouraging and destabilizing. The sculpture is most easily viewed from a distance of about ten feet or so, where it is possible to more fully experience its shape and textual content. The sculpture's height invites the viewer to sit with the message and experience the difficult position of the labor reform activists and Mill Girls who, while strong and persistent, were also, as a fact of their existence, immersed in discontinuity and chaos.

The Fourteen Hour Clock

Just past the *Seating Circle* is the third path marker (Figures 18 and 19). Following this, visitors will come across the last of the large sculptures comprising *Industry Not Servitude.* This last sculpture is perhaps the most prominent and disconcerting of all, consisting of a large granite clock that stands out on the landscape and can be seen from a distance (Figures 24–26).

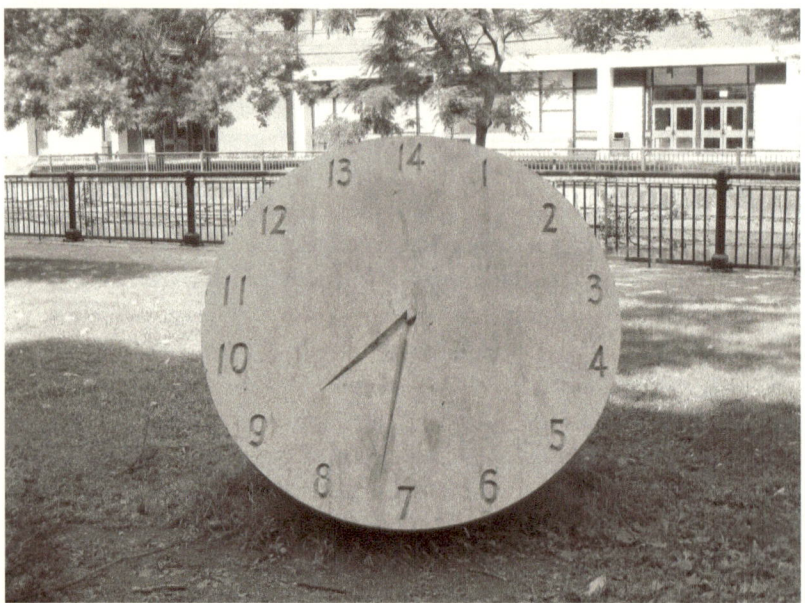

Figure 24: *Fourteen Hour Clock*: Front view. LMNHP. (Photo by the author.)

The clock sculpture commemorates the Ten-Hour Movement, of which Sarah Bagley was the biggest proponent.

The clock stands about five feet tall and at a slight angle, with its large, flat face facing out toward the walking path. When I asked the park ranger which aspects of Lucy Larcom Park she thought most resonated with visitors, she felt visitors were most "struck by the clock." The clock is indeed striking but also destabilizing in its content. It too is composed of granite, affording it a weighty, imposing quality. Upon looking at the clock, visitors will immediately know that something is not quite right but may be unsure of the specific problem they sense. Upon further observation, the visitor will notice that the clock contains fourteen hours, not twelve.

Instead of the typical twelve digits on its face, the clock contains numerals from one through fourteen, indicating the fourteen-hour days typically worked not only by the early Mill Girls, but also by the mill workers of the later immigrant era. The positioning of the clock's minute and hour hands indicate a time that, on a typical clock, would be approximately 7:30; however, part of what makes this clock so destabilizing and almost eerie is that visitors are unable to read its time upon a quick glance. Like the path markers or speed bump sculptures, this sculpture too relies on a prior schema of Western culture—one that assumes an occasional, quick glance at a wristwatch or clock. Thus, the sculpture's power not only rests in its obtrusive size and imposing stature, but also in its ability to throw the viewer off balance, so to speak. It forces a double-take, or even a triple-take, undoing assumptions of how time is understood. And indeed, the Mill Girls understood time differently. They did not have the luxury of the typical eight-hour workday; instead, they worked anywhere from 12 to 14-hour days, as reflected in this sculpture's undoing of traditional assumptions about time.

The clock quite literally figures on the landscape as heterochronous; that is, Foucault notes that heterotopias are "often linked to slices in time," or what he calls *heterochronies* ("Of Other Spaces" 26). Heterochronies may represent indefinitely accumulating time, such as the archived information in museums or libraries, or time in its most fleeting of terms, such as the ephemerality of a concert or performance. The clock as may be viewed as the quintessential example of a heterochronous artifact, in that it not only reflects the indefinitely accumulating time of the 14-hour workday, but also has an immediate impact on the visitor's temporal experience, through its symbolic and material weight on the landscape, and the destabilizing bodily experience it engenders.

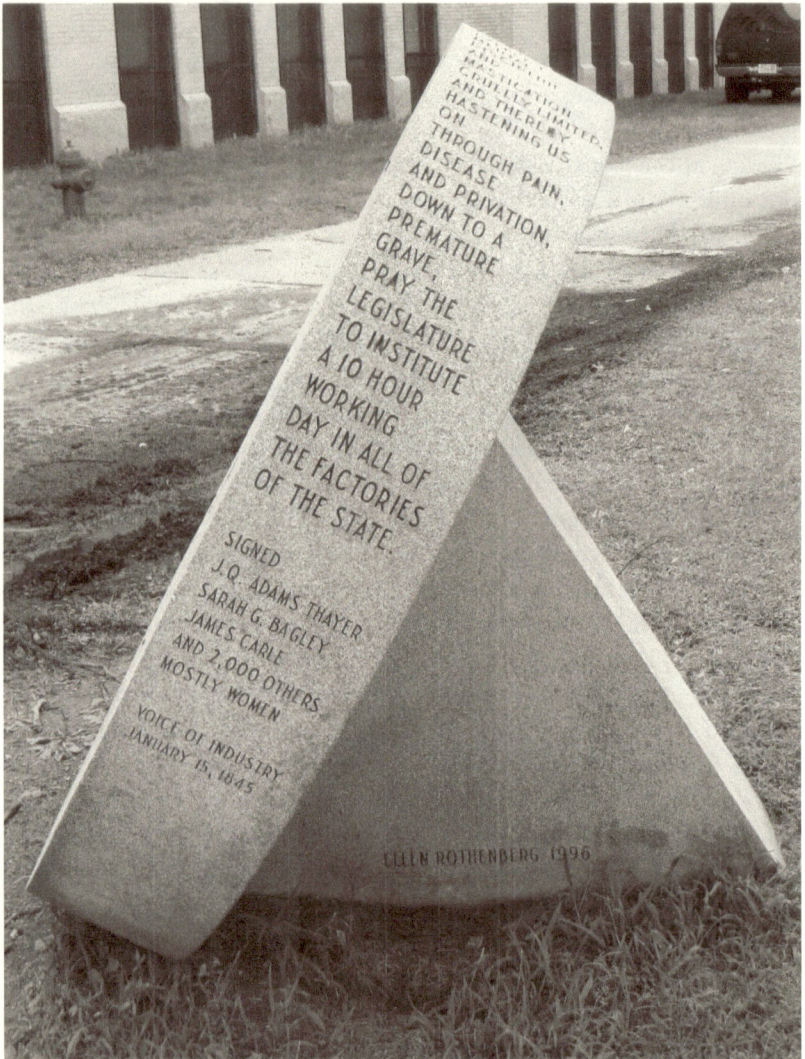

Figure 25: *Fourteen Hour Clock*: Side view with petition inscribed. LMNHP. (Photo by the author.)

In addition to this disconcerting aspect, the sides of this granite clock are also inscribed with the petition made to the Massachusetts legislature for the 10-hour workday. The inscription along the side of the clock reads as follows:

> Petition to Massachusetts legislature
>
> We, the undersigned
> Peaceable, industrious and hardworking men and women of Lowell, in view of our condition, the evils already come upon us, by toiling from 13 to 14 hours per day, confined in unhealthy apartments, exposed to poisonous contagions of air, vegetable, animal and mineral properties, debarred from proper physical exercise, mental discipline, and mastication cruelly limited, and thereby hastening on us through pain, disease, and deprivation, down to a premature grave, pray the legislature to institute a 10 hour working day in all of the factories of the State.
>
> Signed
> J.Q. Adams Thayer
> Sarah G. Bagley
> James Carle
> And 2,000 others
> Mostly women
> *Voice of Industry*
> January 15, 1848
> (Rothenberg, *Fourteen Hour Clock*)

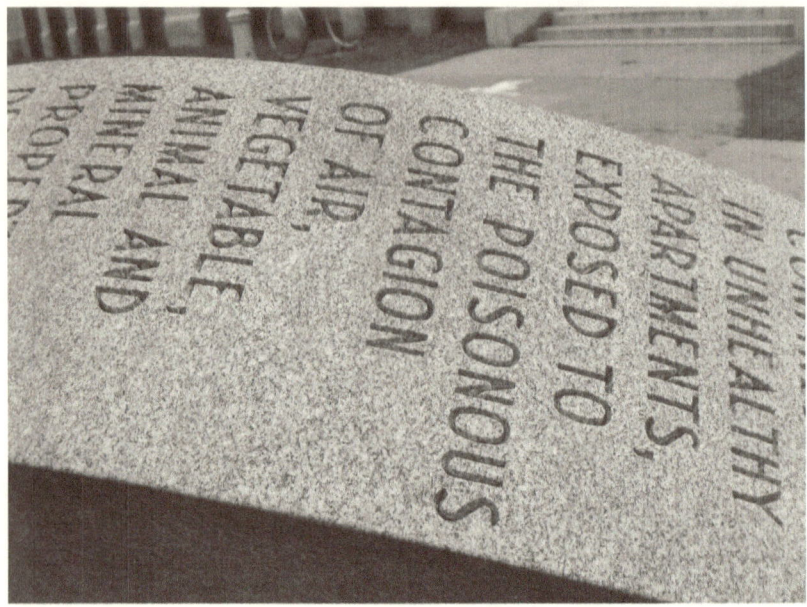

Figure 26: *Fourteen Hour Clock*: Close-up view with inscription. LMNHP. (Photo by the author.)

The granite clock provides a durable canvas for this text, the reading of which is again a physically taxing process; to read the petition in its entirety requires that visitors tilt their heads and walk back and forth, kneeling closer to the ground as they reach its end. The sculpture draws visitors nearer in order to read, but then forces them away in order to fully experience the impact of its size. The sculpture indeed acts on the mind and body; it is physically imposing, induces an overwhelming dizziness through its reading, and is highly destabilizing upon first glance. Even if viewers come to the park with a prior knowledge of Lowell's history and the Ten-Hour Movement, which according to the park ranger, many do, the clock is still striking. No matter what the visitor understands of the Mill Girls' history, their instinct is still to look at the clock and expect to see twelve hours on its face; instead, they are caught off guard and must ask, for just a moment, what is happening here?

Even though the clock ostensibly documents the Ten-Hour Movement and the time period just after the *Lowell Offering* stopped publication, it arguably reflects and performs the crisis heterotopias of the early Lowell Mills—that of a privileged space whose inhabitants share a common dilemma. In this case, the Mill Girls wanted to believe that the mills presented a better way of life than the farms on which they grew up, but in working there, they soon became disenchanted with mill life and the new struggles that it brought—among them, long work hours and a relationship with the mills that was both filled with reverie and a sense of intimidation. Similarly, the clock sculpture is intimidating in its imposing stature and destabilizing content; it is a visually striking addition to the landscape of Lucy Larcom Park, and demands that its viewers stop, decipher, and heed its message.

Walking northeast through Lucy Larcom Park, the clock sculpture is the last of the large sculptures visitors will encounter before leaving the park (Figure 27). The clock is followed by the fourth path marker (Figures 20 and 21): "And you will come off victorious [. . .] We say 'stick to your text.'" The text of this sculpture again suggests that advocates of the Ten-Hour Movement must be persistent in promoting their cause. Moreover, "stick to your text" takes on an additional layer of heterochrony given its inscription into this durable granite sculpture that will indefinitely preserve and promote the message both through its imposing visibility and in the durability of its material.

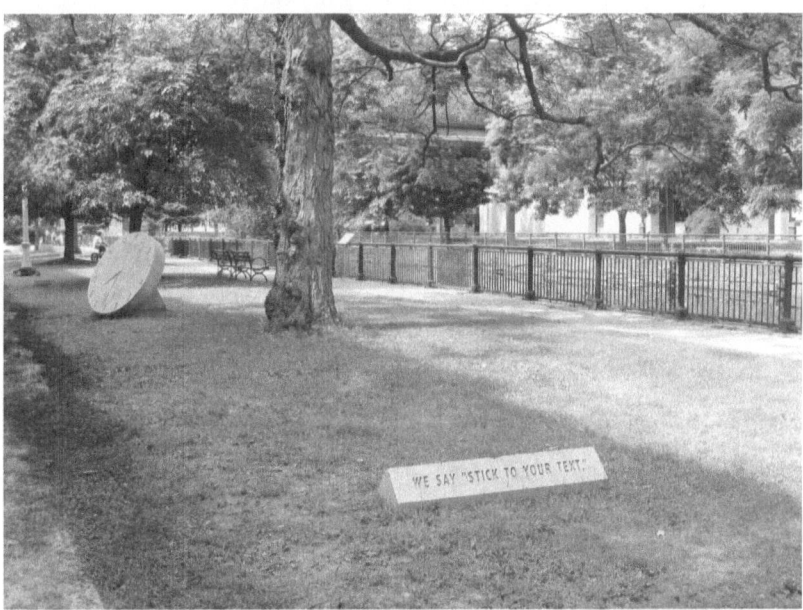

Figure 27: Lucy Larcom Park: From the north, facing southwest, with view of *Fourteen Hour Clock* and last path marker. LMNHP. (Photo by the author.)

Textual content is an important component of the sculptures that comprise the *Industry Not Servitude* installation in Lucy Larcom Park. The material composition and spatial layout of the sculptures, combined with their textual inscriptions, produce a visual-material rhetoric that engages the mind and body of the park visitor, and also reflects and performs the experiences of the early Mill Girls and labor reform activists of the later mill era. By sticking to their text, so to speak, these sculptures have a greater impact on the viewer's understanding of the environment in which the Mill Girls worked, and the labor reform movement that was a direct result of the conditions of that environment.

In leaving Lucy Larcom Park headed northeast, visitors will soon find themselves back in downtown Lowell. To move from the park's green spaces to the sidewalks and intersections of the city center can be destabilizing experience; that is, to be immersed in the park's heterotopic performance at one moment, and to navigate the city streets and traffic patterns in the next, requires visitors to quickly toggle between two states of mind. In this sense, visitors too must shift between the natural and industrial settings in which the Mill Girls lived. Upon exiting Lucy Larcom Park, visitors must cross the street in order to stay on the Canalway walking path. In about one tenth of a mile, visitors will arrive at the Boott Mills complex, which contains Boardinghouse Park.

BOARDINGHOUSE PARK

Boott Mills (Figure 28) is the only fully-preserved cotton mill in the Lowell Mills Park. It has been converted into a museum that houses a room with twenty functioning looms as well as exhibits describing the mills and machinery.[10]

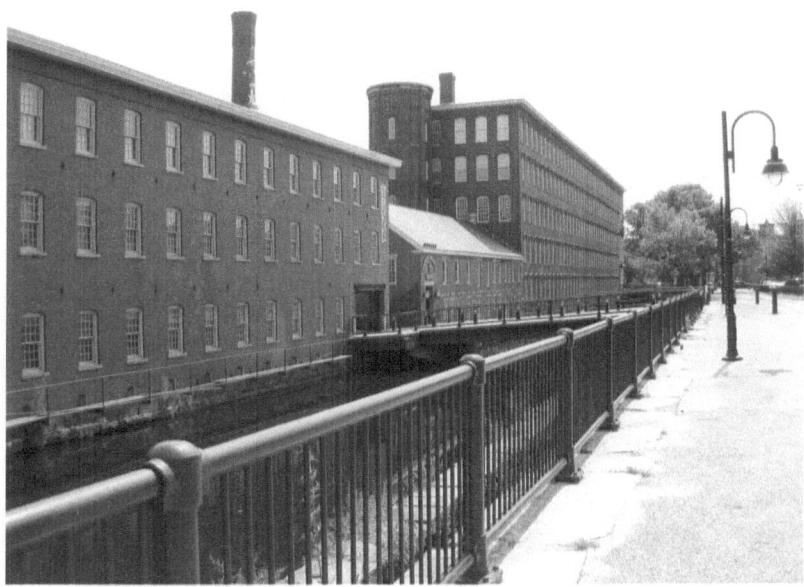

Figure 28: Boott Mills. LMNHP. (Photo by the author.)

Perpendicular to Boott Mills are the Morgan Cultural Center, former-ly a boardinghouse (Figure 30), and the Park Headquarters, formerly the Agent's House (Figure 31). In front of Boott Mills is a large green space that is home to Boardinghouse Park (Figure 29).

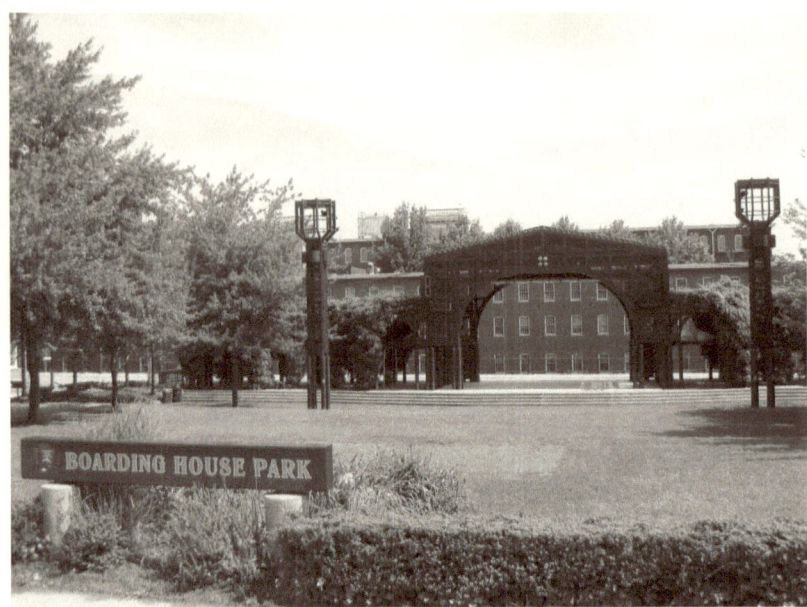

Figure 29: Boardinghouse Park: View with Boott Mills in the background. LMNHP. (Photo by the author.)

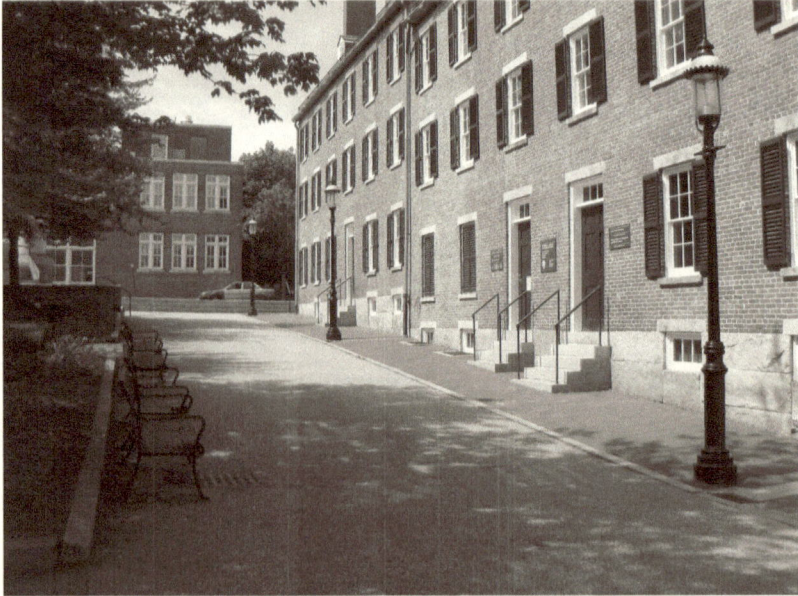

Figure 30: Morgan Cultural Center (formerly a boardinghouse): View with side of Agent's House/ParkHeadquarters visible in background. LMNHP. (Photo by the author.)

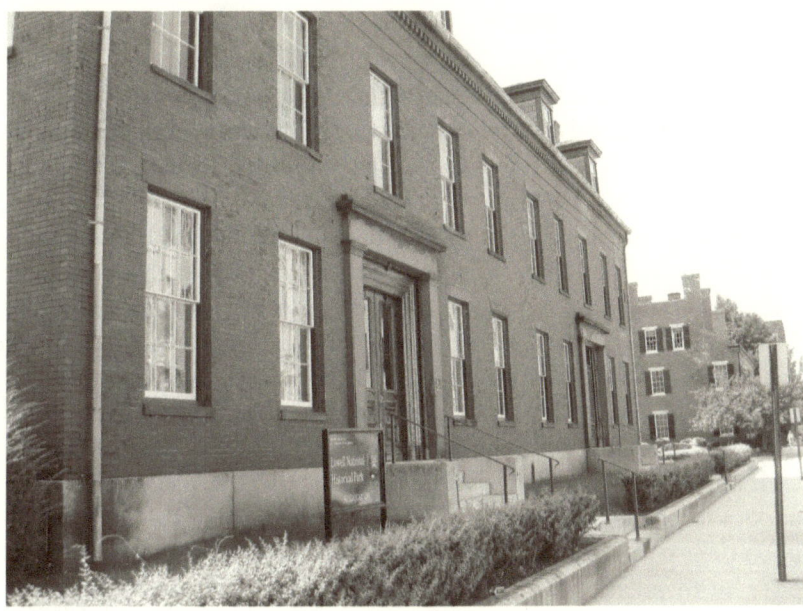

Figure 31: Park Headquarters (formerly the Agent's House). LMNHP. (Photo by the author.)

While I describe the mill and surrounding structures in relation to how they influence the feel and layout of the park, most relevant to this analysis is the park itself, and I therefore delimit my formal analysis solely to Boardinghouse Park, which is comprised of a green space in front of the Boott Mills complex. Boardinghouse Park too performs the crisis heterotopias and heterotopias of compensation in which the Mill Girls lived. The spatial relationships fostered by this park relative to its surrounding structures help reinforce the rhetorical significance of the space of the mills and boardinghouses relative to their impact on the bodies of the Mill Girls.

Boardinghouse Park is both bright and eerie, and very much captures what I can only describe as a sort of collective energy of the Mill Girls. The park is situated in a large green space of about an acre in size. To its north is the Boott Mills Cotton Museum and the Eastern Canal; to its east is the Boott Mills Trolley Stop (which shuttles tourists around the perimeter of the park); to its south is French Street and historic Lowell; and to its west, bordering the park green space, is the Morgan Cultural Center, formerly a boardinghouse. Farther southwest, on the other side of French Street is the Park Headquarters, formerly the Agent's House. This means that Boardinghouse Park is directly bordered on two sides by mill buildings, one of which was originally a boardinghouse, the other of which is the Boott Mills. The park may be likened to a quad on a college campus—a small, square, green space surrounded on two sides by buildings associated with the space.

There are three primary routes that visitors may take to arrive at Boardinghouse Park, and they may either use the Canalway walking path or the Downtown walking route to do so. If visitors take the Downtown walking route, they have the option of entering the park at the front entrance (at the south side, along French Street), or from the east, along John Street. If they take the Canalway walking path, they will enter the park from the north side, closest to Boott Mills. In each of these cases, visitors will be struck by two main elements of the park: a large concert stage in the center of the park, and, depending upon their point of entry, one of three sculptures, which together comprise an installation called simply, *The Lowell Sculptures: One, Two, and Three.*

The Lowell Sculptures: One, Two, and Three

The Lowell Sculptures: One, Two, and Three were created in 1990 by the artist Robert Cumming, and are quite different from the installation in Lucy Larcom Park. Composed of granite, brick, and steel, these sculptures may be understood as resonating with visitors' notions of industrial Lowell, by presenting items or shapes associated with factory life. The plaque (Figure 32) describing the sculptures reads:

> These sculptures, located in three corners of the park, are composed of simple forms based on Lowell symbols that have been combined in a modular design. The shapes represent aspects of factory cities. The thread spool/beehive piece, for example, is a symbol of industry. *Sculpture One* includes a six-ton silhouette of Francis Cabbot Lowell, for whom the city was named. (Lowell Public Art, "The Lowell Sculptures")

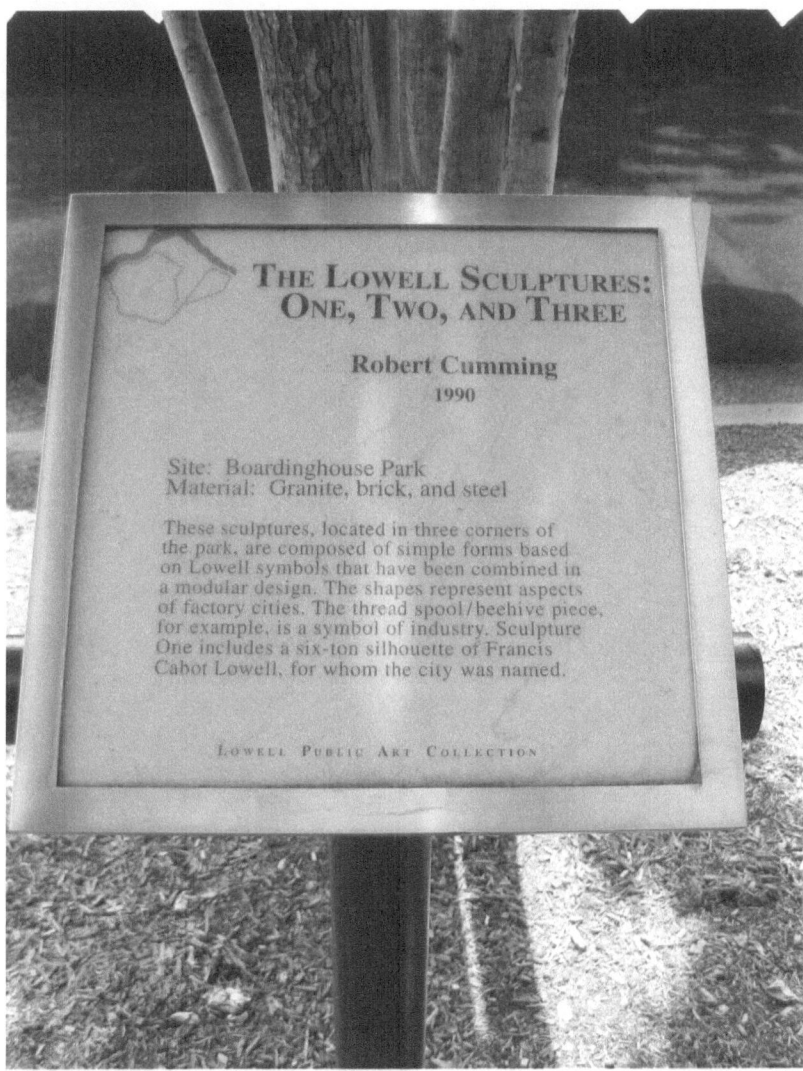

Figure 32: Informational Plaque: *The Lowell Sculptures: One, Two, and Three.* LMNHP. (Photo by the author.)

These three sculptures, as the images show, have a quite different feel from those sculptures in Lucy Larcom Park. They appear taller, positioned atop pedestals, and look dense and cumbersome. They are not inscribed with text or quotations like those in Lucy Larcom Park.

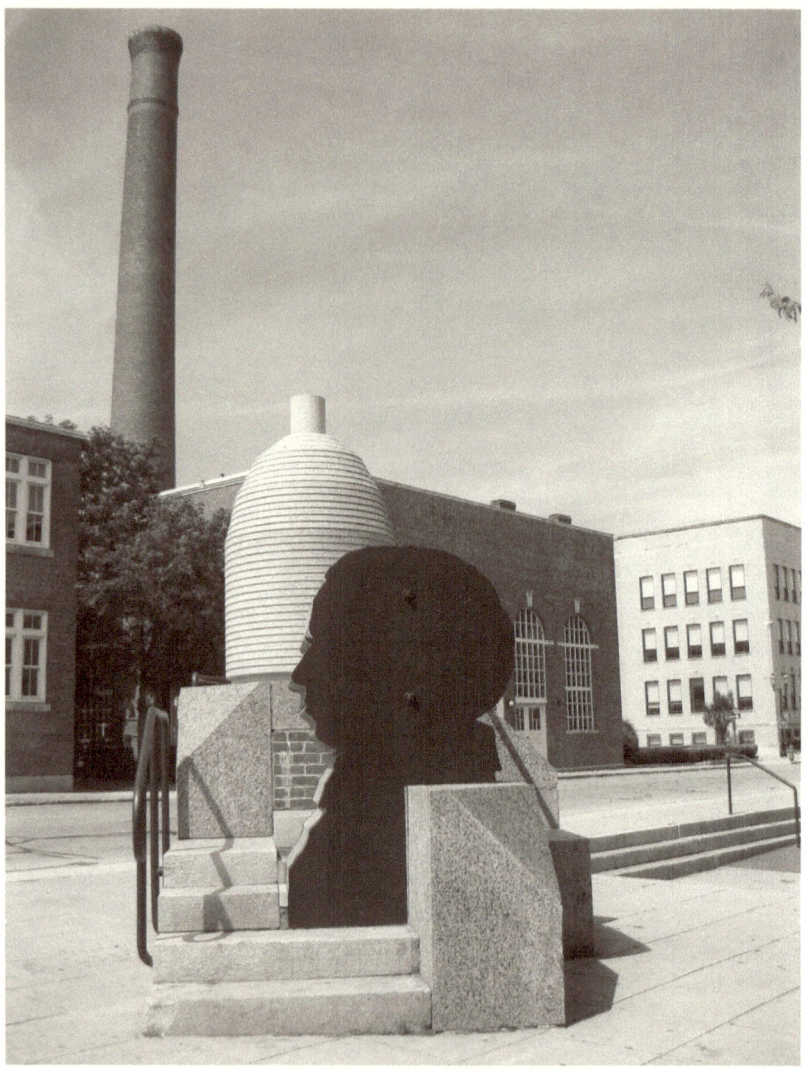

Figure 33: *Sculpture One.* LMNHP. (Photo by the author.)

Sculpture One (Figure 33) is situated at the southwest corner of the park, and would be the first sculpture encountered by visitors walking east on French Street, along the Downtown walking route. To visitors, the sculpture might appear rather austere and formal-looking, particularly because it does not quite mesh with the ineffable, almost eerie quality of the park itself or the concert stage that it borders. Again, though, it is impossible to truly know the artist or author's intent in the design and placement of this sculpture, and so it is only possible to describe the consequence of the piece from the vantage point of the visitor or observer. The sculpture has two levels and again invokes the theme of ascending stairs, with a small set of seven steps that visitors can ascend. With the inclusion of seven steps and a prominent railing, this sculpture invites its viewer to engage with and traverse its stairs more so than the *Steps* sculpture in Lucy Larcom Park. Toward the base of the sculpture, approximately three steps up, is the silhouette of Francis Cabbot Lowell, to which the informative plaque refers. Atop the steps, situated on a brick platform, is a large, granite, beehive-shaped thread-spool; thus, the park rangers informally refer to this piece as the Beehive Sculpture. The beehive metaphor invokes both the idea of the worker and also the honeycomb structure of the hive, perhaps representative of the tight living quarters shared by the Mill Girls.

Sculpture Two (Figure 34) sits at the southeast entrance of the park and would be encountered first by visitors walking north on John Street, along the Downtown walking route.

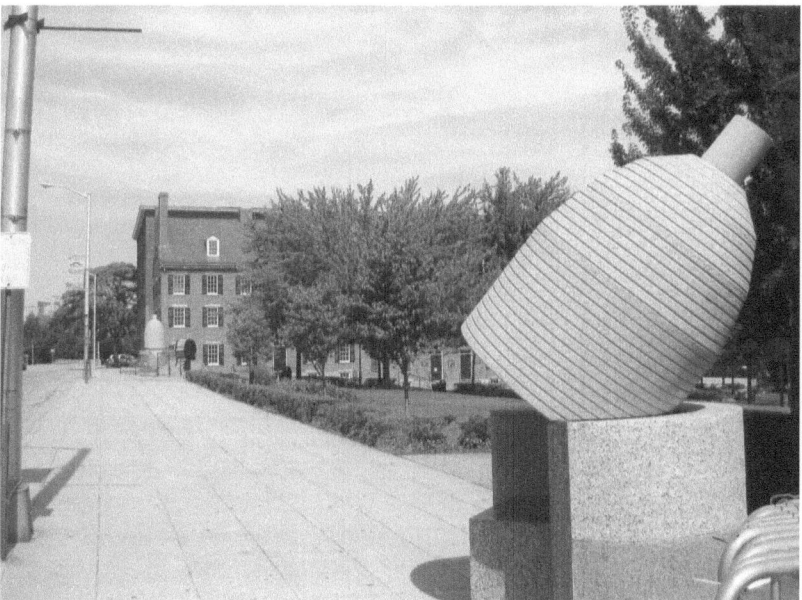

Figure 34: *Sculpture Two*: View with *Sculpture One* facing northwest on French Street. LMNHP. (Photo by the author.)

It too is a beehive-shaped thread-spool, and sits atop a granite plat-form. The sculpture is about eight or nine feet tall including the base, and visitors may again notice its dense, cumbersome, overbearing, and austere quality. While it is said to represent factory cities in general, because the sculpture sits at the edge of Boardinghouse Park, it may also be read as symbolic of the living conditions endured by the Mill Girls specifically. Again, to experience the sculpture is to understand the symbiotic work of materiality and symbolicity (Blair 19). Like *Sculpture One*, the beehive again invokes the idea of laborer; it also invokes the idea of the honeycomb, which may be understood as meta-phor for the tight living quarters in which the Mill Girls lived.

Sculpture Three (Figure 35) sits at the northeast entrance of the park, and would be the first sculpture encountered by anyone ap-proaching Boardinghouse Park by way of the Canalway walking path, along the north side of the park.

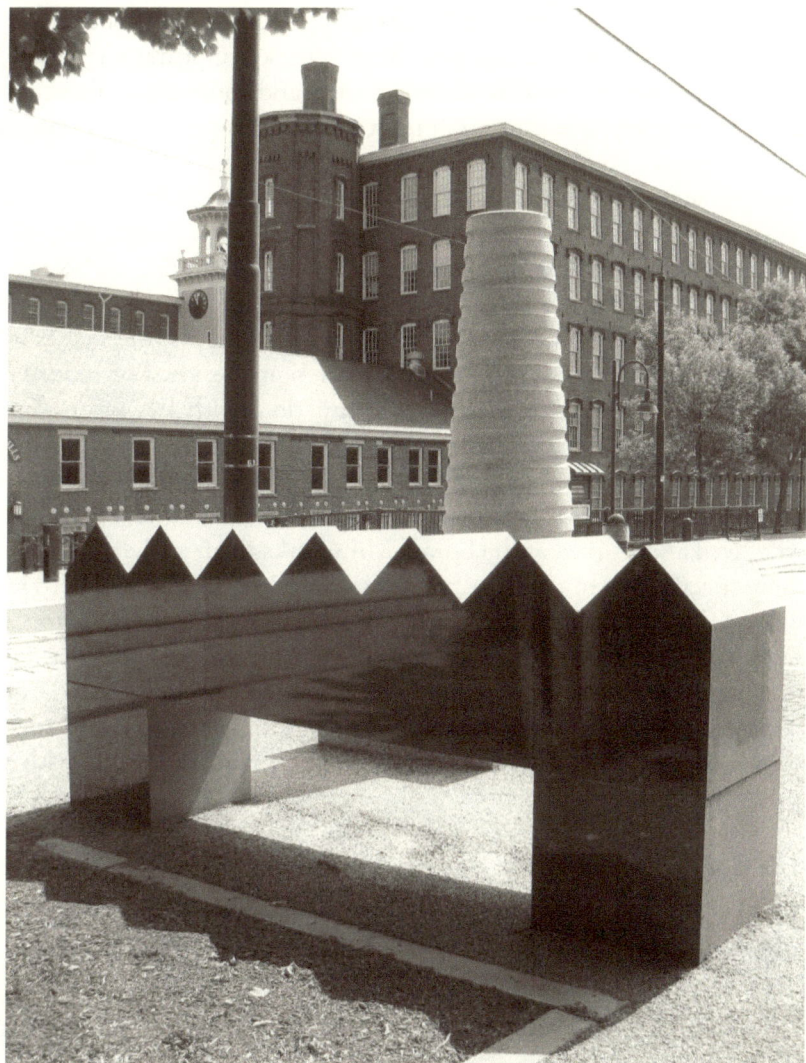

Figure 35: *Sculpture Three.* LMNHP. (Photo by the author.)

It is composed of a tall, narrow, spool-like structure set next to a jagged, polished granite object that is shaped like a bridge, whose top has a series of pointed, triangular shaped edges. Presumably representative of a factory city, it likely represents the machinery of the textile mills.

These three sculptures do not interrupt the landscape or demand the sort of taxing physical engagement as do those in Lucy Larcom Park; their observation and interpretation do not require or invite the same type of physical and emotional engagement as those sculptures comprising *Industry Not Servitude*. Rather, these sculptures serve more so as material markers that delineate the park borders and function as a symbolic testament to the impact of the mills on Lowell's industrial heritage. They are visually imposing on the landscape but are situated such that they do not interrupt the visitor's physical walking space as they proceed along the walking tour. Interestingly, they almost seem to guard Boardinghouse Park while delineating its boundaries. In doing so, the three sculptures help reinforce the notion of the park as heterotopic space. On the one hand, that is, the shapes of the sculptures themselves are said to represent the ideas of industry and factory cities generally, and thus function, with less particularity, "outside of all places" (Foucault, "Of Other Spaces" 24). On the other hand, that these sculptures physically punctuate the boundaries of Boardinghouse Park, a park that serves to commemorate the women who lived and worked at the Lowell Mills, links them to the Mill Girls more specifically. In this case, their imposing size, height, and heft, as well as the placing of the silhouette of Francis Cabbot Lowell atop a set of stairs, may be read as representing the hierarchical relationship between the forms and shapes of Lowell and their immense weight or impact on the bodies of the Mill Girls; given this reading, the sculptures also illustrate heterotopic space through their relationship with a particular "location in reality" (24).

The Concert Stage

The concert stage in Boardinghouse Park was built between 1989 and 1990 and often serves as the setting for the Lowell Folk Festival and other venues (Park Ranger).

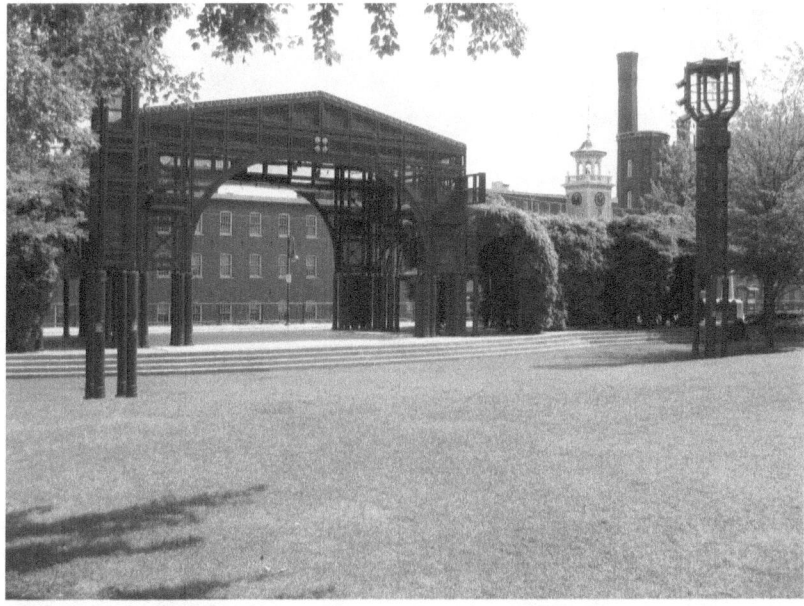

Figure 36: Concert Stage in Boardinghouse Park: Front view. LMNHP. (Photo by the author.)

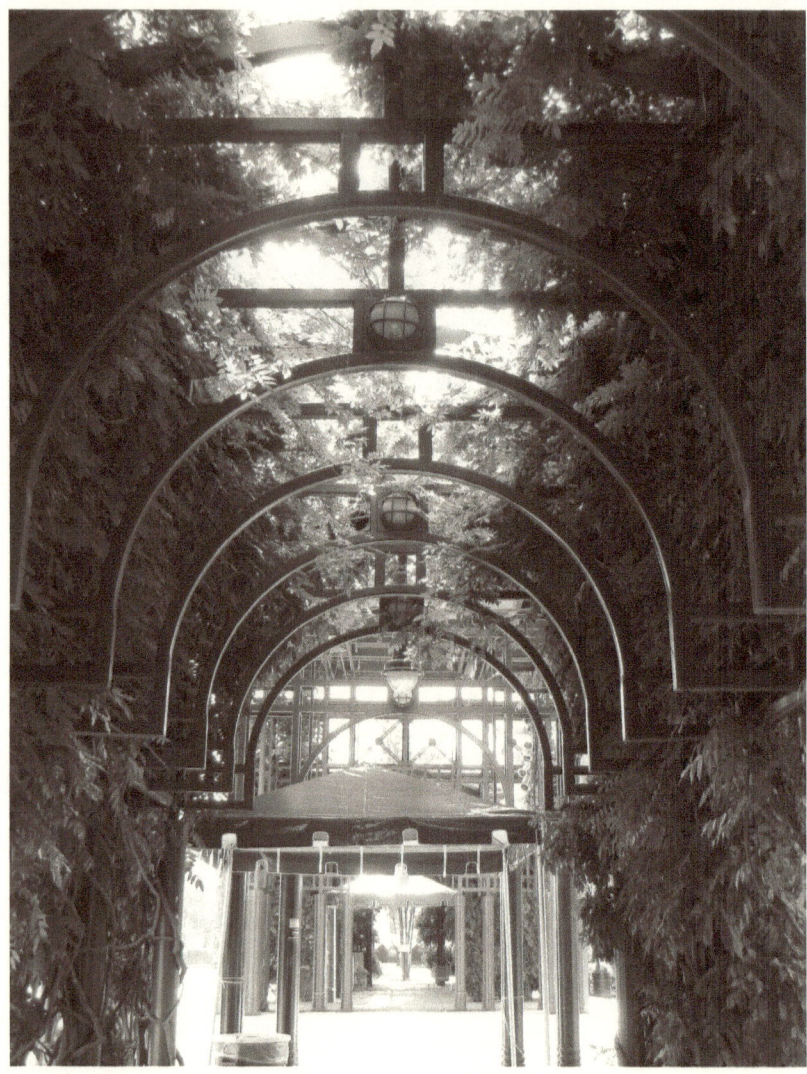

Figure 37: Concert Stage in Boardinghouse Park: Inside view. LMNHP. (Photo by the author.)

The concert stage is situated in the middle of the green space and set against the backdrop of the Boott Cotton Mill (Figures 36 and 37). As an object of visual-material rhetoric, the stage has a beautifully wistful ambiance, coupled with a remote eeriness. The stage is a metal construction, and with its dark green color and series of high arches, was meant "to fit into the landscape," [11] perhaps performing once again the mill owners' attempts to merge industry with the natural environment. (Park Ranger).

The concert stage itself is also built over what were originally two boardinghouses, and in this sense, may be understood as commemorating the lives of the Mill Girls. As both the site of former boardinghouses and the present-day venue for theater performances and musical events such as the annual Lowell Folk Festival, the stage is perhaps the quintessential heterochrony in its embodiment of multiple, shifting temporalities. Foucault describes events such as the festival as embodying "time in its most fleeting, transitory, precarious aspect" ("Of Other Spaces" 26). Boardinghouse Park then represents both the timeliness of the "heterotopia of the festival," such as that of the Lowell Folk Festival, and "that of the eternity of accumulating time," invoked by the idea of the Mill Girls' timeless presence at Boardinghouse Park (26). Understanding the work of Boardinghouse Park not only as heterochronous but also as visually and materially rhetorical can also help account for how bodies engage and are engaged by this physical space.

Because Boardinghouse Park is not a linear stretch of green space like Lucy Larcom Park, visitors are not guided through the site in quite the same way. Instead, they make inferences about the space and its history through a combination of historical contexts articulated by park plaques and through visual, spatial comparisons of the surrounding built structures. The park ranger felt that visitors tend to notice the former boardinghouse along the west edge of the park and its close proximity to both the Boott Mills and the Agent's House across the street; in observing the juxtaposition of these structures, visitors are able to make observations about the mill workers' lives. They are struck, for example, by the way in which the mill operatives "lived next to their work" (Park Ranger). Because they are also able to see that the boardinghouse (now the Morgan Cultural Center) is next to the Agent's House (now the Park Headquarters), they begin to make connections about "the unfairness of these living conditions, because the Agent's House was home only to the agent's family, but is about the same size as a boardinghouse, which housed up to 250 women." Thus the visual and cultural landscape of the park "drives home the tension between emerging classes" (Park Ranger). In this way, Boardinghouse Park may serve as a memory text that moves visitors to infer political ideas about the lives of the Mill Girls; that is, the park's rhetorical work may be an "effect of what and how we remember, and the uses to which those memories are put" (Biesecker, "Remembering" 168–169). Indeed, the juxtaposition of the Agent's House with the former boardinghouse next to the park may allow visitors to see that, while the two buildings are approximately the same size, one was home to a small family while the other was home to nearly 250 women. Boardinghouse Park may be understood then as subtly advocating for the vulnerable bodies of the Mill Girls by maintaining the visceral dimensions of their lived experience through a combination of spatial and textual cues present in the park.

The boardinghouse also looks small in comparison with the Boott Mills behind it. A plaque near the concert stage describes the boardinghouse and implicitly makes this point. The plaque is titled, "In the Shadow of the Mills," and its main paragraph reads: "To the right stands a boardinghouse block built in 1837 for Boott Cotton Mills workers. Dozens of company-owned boardinghouses served as home for thousands of young, single women—Lowell's 'mill girls'" ("In the Shadow"). The plaque's title, "In the Shadow of the Mills," addresses the spatial relationship between the mill (which housed the huge machinery operated by the Mill Girls), and the boardinghouse (which was much smaller and afforded only small, cramped spaces in which the Mill Girls ate and slept). The plaque's main paragraph in itself functions as a spatial narrative, which in its description of the ratio of "thousands of Mill Girls" to "dozens" of boardinghouses, also alludes to the cramped living quarters of the mill operatives. This information, coupled with the visual disparity of the boardinghouse juxtaposed with the mills and the Agent's House across the street, provides visitors with both the visual and spatial information necessary to make inferences about lives of the Mill Girls and the emergence of the classes that came with the industrial revolution.

The power of Boardinghouse Park and its artifacts thus lies in its ability to illuminate the spatial relationships that acted on minds and bodies of the Mill Girls and ultimately help visitors draw conclusions about the implications of those living conditions. The concert stage also invites visitors to a space where they can be exposed to artifacts that promote an awareness of the Mill Girls' history. Thus Boardinghouse Park, as a visual-material rhetoric of heterotopic space, allows visitors to engage with greater empathy in the experiences of the Mill Girls.

HOMAGE TO WOMEN

While Boardinghouse Park certainly does well to commemorate the Mill Girls and specifically promotes a better understanding of their difficult living conditions in the boardinghouses of the Lowell Mills, perhaps the Mill Girls are most overtly and distinctly memorialized in Mico Kaufman's 1984 sculpture, *Homage to Women* (Figures 38–44).

Homage to Women, composed of bronze and granite, serves to commemorate the Mill Girls; additionally, as its interpretive plaque states, its "intertwined figures also represent the struggles and aspirations of all women throughout time" (Lowell Public Art, "Homage to Women").

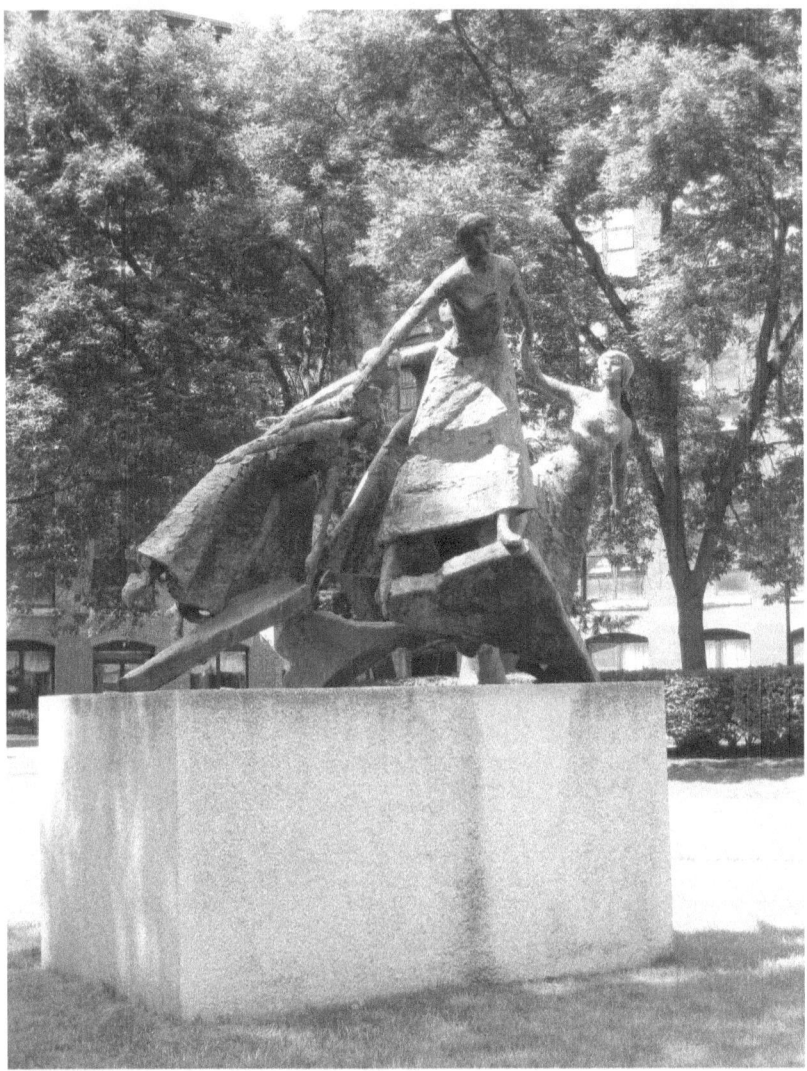

Figure 38: *Homage to Women.* LMNHP. (Photo by the author.)

The sculpture is often included as a stop along tours given by the park guides—especially those guides who focus on what is informally called the "Mill Girls Tour" (Park Ranger). Like the installations in Lucy Larcom and Boardinghouse Parks, *Homage to Women* is considered Lowell Public Art, and the site of the sculpture is likewise demarcated with a brown circle on the park map. The sculpture resides in a small, enclosed green space just behind the park's visitor center (Figure

39). This park has a quiet peacefulness about it, and because it is bordered by tall buildings and is situated behind the visitor center, it feels protected and ensconced. Four entrances paved with brick walkways guide visitors into this small green space and lead visitors to the sculpture, which sits just off-center from the middle of the site.

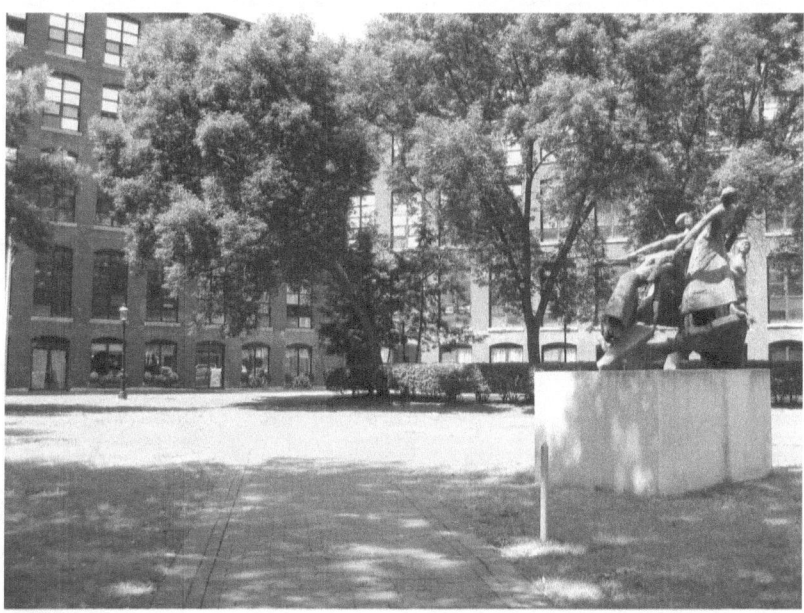

Figure 39: *Homage to Women*: View of surrounding green space. LMNHP. (Photo by the author.)

Park benches also line the perimeter of the area. On the north side of the park is Market Street, which runs parallel to the Canalway walking path. To the west of the site is the main visitor center and Market Mills, a renovated condominium complex and office building. Just east and south of the area are the National Park limits, beyond which extend the city of Lowell and the Pawtucket Canal, which means that this memorial really is on the outskirts of the historic park, if not just beyond its borders. Nonetheless, the memorial does reside just behind the visitor center and is quite accessible.

Homage to Women helps emphasize the role of the Mill Girls in Lowell's industrial history and helps show that the Mill Girls are indeed worth studying. As Blair has put it, memorials have an "agenda-setting function"; in other words, she says, "when a memorial [. . .]

appears on the landscape, it is thereby deemed—at least for some, and at least for the moment—attention worthy" (35–36). The fact of the memorial's existence affords it the chance to garner attention that it wouldn't otherwise have (36).

The sculpture itself stands about ten feet tall, including its base. Made of granite and bronze, it is obviously made to endure the elements in ways that text on paper is not. In its durability, the sculpture performs the persistence requested by the path markers in Lucy Larcom Park. The sculpture is made of bronze, which, while able to withstand the elements, is also difficult to photograph. This difficulty in conveying the nuanced facial expressions of the women in the sculpture thus reveals the limitations of the reproduction as an "intervention in the materiality of the text" (Blair 38). To experience the sculpture firsthand is therefore a very different experience than to view its reproduction on the park's web site, or even in the photographs reproduced in this chapter. To view the sculpture in person is to more fully experience the postures and facial expressions of these five women and to imagine the possibilities for how their movements and gestures may be interpreted.

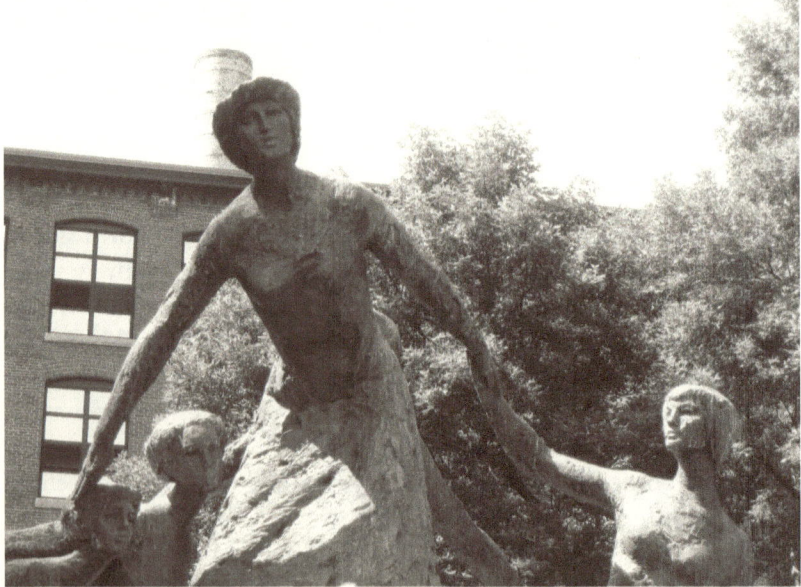

Figure 40: *Homage to Women*: Looking up. LMNHP. (Photo by the author.)

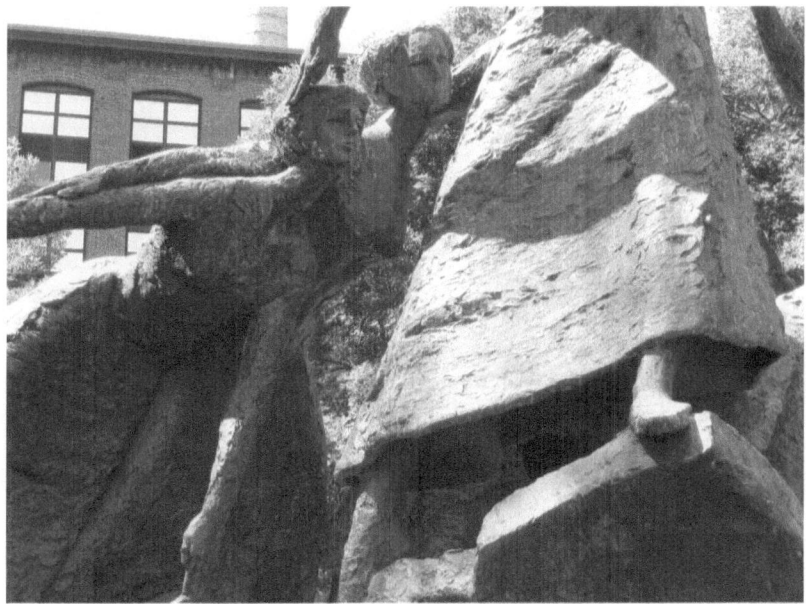

Figure 41: *Homage to Women*: Close-up 1. LMNHP. (Photo by the author.)

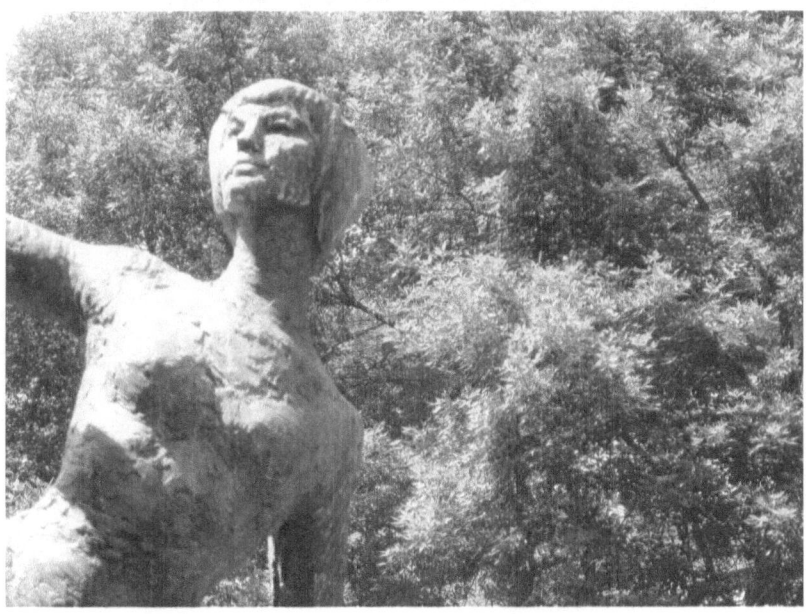

Figure 42: *Homage to Women*: Close-up 2. LMNHP. (Photo by the author.)

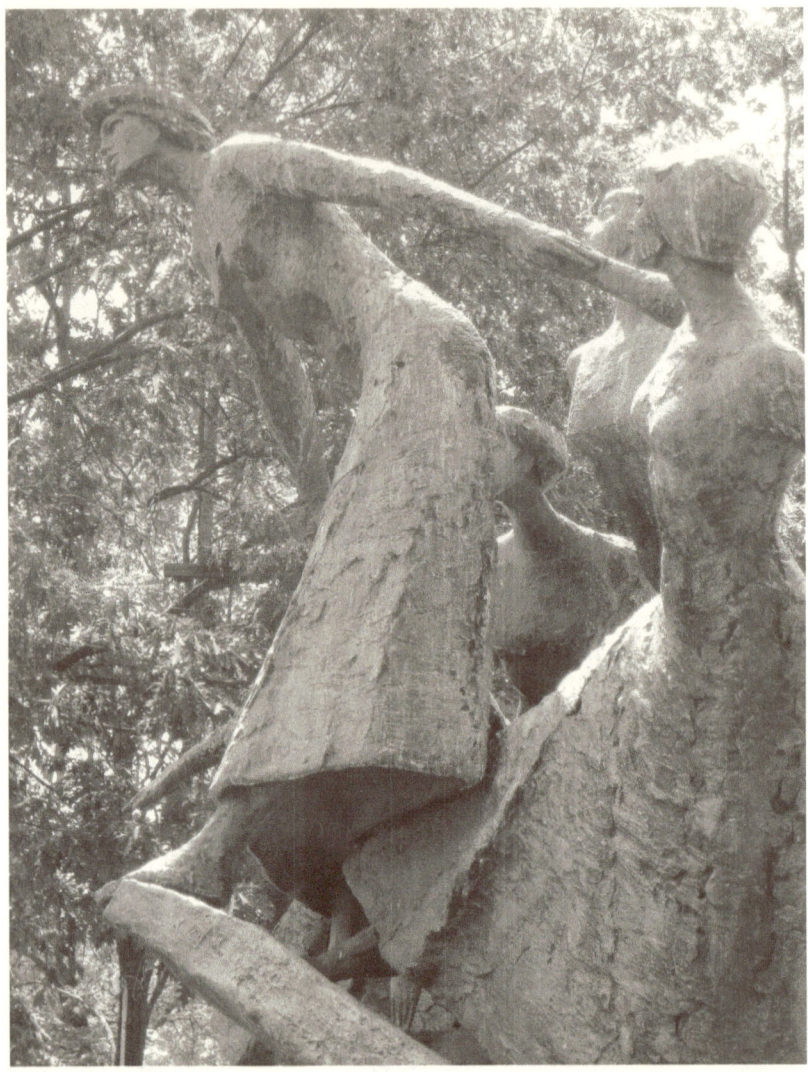

Figure 43: *Homage to Women*: Close-up 3. LMNHP. (Photo by the author.)

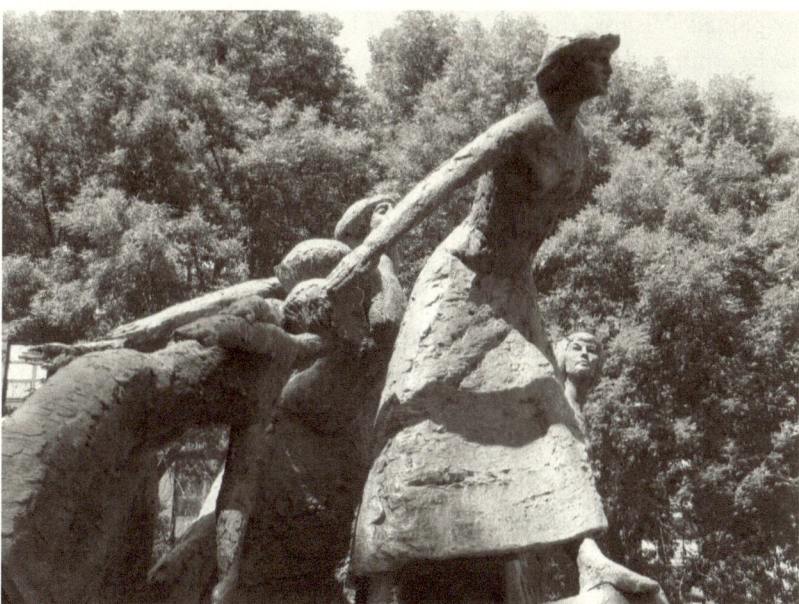

Figure 44: *Homage to Women*: Close-up 4. LMNHP. (Photo by the author.)

Just as the Mill Girls' narratives speak expressively to the reader through the pages of the *Lowell Offering,* the five Mill Girls in *Homage to Women* speak to their viewer through their gestures and expressions. Visitors' interpretations of this sculpture often challenge one another, and as the park ranger has put it, no singular interpretation is necessarily the correct one.

To consider the possible interpretations of *Homage to Women* is to engage with this material text in such ways that it controls the "vector, speed, or possibilities of physical movement" (Blair 46). The sculpture is meant to "act on the whole person, not just on the 'hearts and minds' of its audience" (46). Because of the sculpture's height, and because each of the five women look out in different directions, it is necessary that visitors not only look up at the memorial, but also walk its circumference, zooming in and out on foot, in order to get a holistic experience of the piece.

Not only does the sculpture act on visitors' bodies, inviting, or even requiring them to engage with it, but the sculpture also quite literally *depicts* the Mill Girls' own bodies. Visitors bring their own contextualized, embodied experience to their readings of this sculpture. To visitors, the faces of these five women may appear haunted, concerned, or pained. Visitors may notice that at least four of the women appear to be squinting, that the woman in the front appears to be climbing or balancing on the others, or that the women seem to be clinging to one another. In describing visitors' responses to the sculpture, the park ranger noted that reactions tend to vary widely. For example, some visitors find the sculpture to be "ethereal," while others have said it looks like the women are "leaning into the future." One visitor thought the sculpture represented "leaping off points to a new adventure," while yet another thought the women looked like they were "ready for something different" (Park Ranger). These varied interpretations, coupled with the description provided in the sculpture's accompanying plaque and the multiple readings invited by the sculpture's visuality, physicality, and surrounding park contexts, also illustrate a point made by Marback that "meaning and significance are in the moment, in activities of interacting with others with and through objects" (52).

Like many of the sculptures comprising the art installations in Lucy Larcom and Boardinghouse Park, *Homage to Women,* too, acts on the minds and bodies of park visitors, setting in motion a constellation of contexts and objects that allow visitors to engage with greater empathy with the lives of the Mill Girls. The sculpture functions as a heterochronous, visual-material text that embodies the Mill Girls. Their facial expressions, bodily gestures, and direction of movement in this sculpture perform a material rhetorical account of their lives and struggles. The sculpture's material existence invokes and enacts a timeless symbolicity that deems the Mill Girls' history attention-worthy and helps keep them alive as part of Lowell's contemporary historical landscape.

CONCLUSION

The map, green spaces, and public art installations of the Lowell Mills National Historic Park act on the minds and bodies of park visitors in ways that reflect the crisis heterotopias and heterotopias of compensation of the early Lowell Mills. The park map acts on the body through its selectivity, directing visitors' movements and guiding their navigational choices. The map communicates a point of view about which park features are to be deemed attention-worthy, thus affording it with its rhetorical power.

The five elements comprising the *Industry Not Servitude* installation in Lucy Larcom Park perform both the crisis heterotopias and the heterotopias of compensation of the early Lowell Mills. The *Circular Fence with Poem, Seating Circle,* and *Fourteen Hour Clock* each act on the mind and body in their own way, often disorienting the visitor while inviting them to empathize with the chaos and dissonance experienced by the Mill Girls. The path markers both interrupt the linearity of the park and weave the larger sculptures into an intertextual, visual- material narrative. Through their design and placement, they force the visitor to stop and look; through their textual inscription, they reinforce a message that not only conveys the gravity of the Mill Girls' oppressive environment but also does so through a parallel and chiasmatic structure that works with and against the park's linearity and the placement of the larger sculptures, creating contexts that may allow visitors to discern at least some of the park's story.

The Lowell Sculptures: One, Two, and Three, which line the edges of Boardinghouse Park, do not act on the body in quite the same manner as those sculptures in Lucy Larcom Park, though they do help reinforce the notion of the park as heterotopic space. The sculptures use historic shapes and forms to symbolize the life and work of both mill workers generally and the Lowell Mill Girls specifically. In this sense, the sculptures may be understood as reflecting the crisis heterotopias of the Lowell Mills. The concert stage in Boardinghouse Park creates an eerie heterochrony that commemorates the Mill Girls, not only through its meshing with the natural environment but also by inviting contemporary musical and theater performances. These performances may be interpreted as celebrating the lives of the Mill Girls on a stage built over the former site of two boardinghouses. The juxtaposition of the former boardinghouse along the west edge of the park with the Agent's House across the street helps emphasize the tension between the two groups and the fact of the Mill Girls' unequal living conditions. The park is unique in its ability to construct and perform spatial relationships that act on bodies in the present to ultimately help visitors draw conclusions about the Mill Girls' living conditions and bodily experiences of the past, thus keeping their memory alive and open to continuous contextualized experience.

Finally, *Homage to Women* is consistent with the tone of the Mill Girls' narratives, which often describe a nostalgia for the past and a desire for more livable working conditions. The sculpture encourages visitors to physically engage with the piece and take in the women's facial expressions, gestures, and perceived thoughts, thus inviting varied interpretations that seem to settle on the common theme of hope in the face of continued oppression.

The spatial layout of the park as a whole also communicates the difficulty of the Mill Girls' work environment and the subtle surveillance in which they were immersed. While the park focuses on many aspects of the industrial revolution in early New England, it indeed fosters a destabilizing contemporary account that not only performs the lives of the Mill Girls specifically, but in doing so, engages visitors in ways that embody the overwhelming nature of the Mill Girls' physical environment. Upon initial consideration, the relationship fostered between the park and its visitors might not seem to perpetuate the sort of embodiment called for by Hayles, which focuses on an understanding of how individual, physical, contexualized bodies "impose, incorporate, and resist incorporation" of material practices (*Posthuman* 194). In fact, the park's artifacts and green spaces seem to impose themselves on the visitor in much the same way that the physical environment of the mills imposed itself on the bodies of the Mill Girls. In this sense, the work accomplished by the park and its artifacts might be read as working against a more empowered notion of how subjects interact with their material environments. On the other hand, visitors may resist these impositions by developing their own modes of interaction with the park or choosing how to navigate its artifacts, just as the Mill Girls resisted the oppressive environments of the mills through their own writings and protests. The park then advocates for the Mill Girls by inviting visitors to learn more about the daily lives of these women through what is a challenging but often rewarding engagement with its green spaces, artifacts, and structures. The park thus weaves a complex narrative of visual-material rhetoric that both resists and submits to the oppressive conditions in which the Mill Girls once lived and worked. While the park and its green spaces and artifacts certainly tap into and articulate the historical and symbolic contexts underpinning their presence on the landscape, such contexts and symbolicity hardly constitute the whole of their consequence. Understanding the green spaces and sculptures within the park as visual-material rhetorics of heterotopic space helps support an embodied rhetoric of the mills that emphasizes how the park engages visitors' bodies through its practices of visuality and materiality.

A visual-material rhetorical approach is also valuable in helping to illuminate the ways in which the park advocates for and gives voice to the Mill Girls. A combined Blairian-Foucauldian interpretive lens provides a means of examining how the park's artifacts and spaces work to perform for visitors the embodied experiences of the lives of the Mill Girls. Such an approach to visual-material rhetorics helps show how the park serves to advocate for the Mill Girls by fostering an environment that allows visitors more room to empathize with their struggles. Visitors' physical, embodied experiences within the park in many ways reflect the daily experiences of the Mill Girls while working and living in the mills and boardinghouses. The visual-material rhetoric of the park works in conjunction with other components of the park's symbolicity and historical contexts to provide a more complete rhetorical picture that advocates for marginalized bodies or tangential groups, in this case, the Mill Girls, by fostering a closer bond between the Mill Girls and park visitors. This ability of visual-material rhetorics to help foster greater empathy for tangential groups thus begins to broaden rhetoric's analytical purview from that of its immediacy to its potential for consequence beyond the site of initial goal fulfillment.

This chapter has examined the influence of visual-material rhetorics on contextualized, bodily experience by analyzing the maps, way finding devices, green spaces, and public commemorative sculptures at the Lowell Mills National Historic Park in Lowell, Massachusetts. Subsequently, it demonstrated how these artifacts reflect and perform for visitors the impact of the mills on the lives of the Mill Girls who labored there in the early 1800s. The chapter has helped show how physical, discursive spaces and the artifacts within them can engage bodies in particular contexts. For the most part, visitors to the LMNHP make use of the various maps and wayfinding devices to guide their journey within a specific space with designated parameters. The maps and navigational devices exist to help visitors navigate what is, essentially, a closed course—one that has many possible routes, but that is nonetheless a site delimited by its maps, wayfinding devices, and other prescribed pathways and spatially-delineated boundaries. A visual-material rhetoric of the park helps uncover the various significant impacts of these spaces for the bodies residing within it.

Visual-material rhetorics can also help account for the corporeal impact of navigational experiences through spaces not specifically delineated as are parks, museums, or other interpretive sites. In the next chapter, I consider the ways in which visual-material rhetorics can help account for the contextualized, everyday navigational experiences of technologically-mediated bodies. I consider the ways in which mediated, posthuman bodies work with and against the knowledge claims made by in-car GPS devices to foster spatial understanding and embodied geographic knowledge in the context of everyday, personal navigation.

4 Navigating the Mediated, Posthuman Body

"I'm used to having a map which shows me a guided path to my destination and allows me a view of my surrounding areas, and that was sort of comforting. With the GPS I realized I would not have that, so that was a little concern and remains a little bit of a concern with the GPS" (GPS user OSA9).[1]

"I liked just the map function and being able to see where I was in relation to the surrounding countryside" (GPS user BA14).

"The one time I used [the GPS] for a long distance thing where I kind of knew where I was going, it was telling me to go a way that I wasn't at all familiar with, and I didn't—I didn't do it—it kept recalculating, because I kept going the way I thought it was" (GPS user BA16).

"I love how it suggests an alternate route. I loved how it would recalculate and adapt to me—I was in control" (GPS user BA21).

"I quickly became impressed with how much simpler it was making my travel plans, as well as how much less stress it implied for myself and/or my navigator, since missed exits, wrong turns, and road closings were no longer potentially catastrophic failures" (GPS user OFM10).

As the quotes from these GPS users describe, in-car navigational devices have various, but clearly distinctive impacts on the personal experiences of everyday navigation. Through the combination of their physicality, their visual and auditory cues, and their resulting digital cartographic texts, GPS devices provide multimodal texts for the user. Users may program their routes and then negotiate and reconcile the information provided by the GPS with their own knowledge or interests, ultimately making navigational decisions that impact both the mind and body. Users may work with and against the GPS, they may resist or subvert it, or they may supplement its directives. These interactions between the GPS and the user have the capacity to encourage the development of new geographic knowledge or to constrain or limit spatial understanding. I argue in this chapter that the GPS is a visual-material rhetorical artifact that helps mediate navigational experiences. GPS users then work both with and against the technology and their physical surroundings to make what Kim et al. refer to as "purposeful decisions" (341–342). These purposeful decisions have implications for how we might understand the relationship and interactions between the GPS and the mediated, posthuman body. When referring to the posthuman body, I subscribe to Hayles' (1996) and Brooke's (2000) general positioning of the posthuman as necessarily mindful of the practices of embodiment, and sensitive to the idea that modes of being are inextricably linked to the material world. Given the decision-making and knowledge-making practices fostered through users' interactions with this technology, I argue in this chapter that the GPS and its user co-construct agency, in what Krista Kennedy has called an "interactive process that involves exchanges between multiple agents, texts, and influences" (308). I continue to understand the map as a heterotopic space that can illuminate spatial relationships as rhetorical; subsequently, I show how a visual-material rhetorical approach can help reveal the contextualized processes that shape these levels of interaction and decision-making and their implications for embodied, mediated experience. As a result, I further demonstrate the ways in which it is possible to understand visual-material rhetorics as a sustainable project of inquiry—one that can function across contexts and objects of analysis.

The ability of visual-material rhetorics to account for posthuman, mediated modes of understanding resides once again in its interest in understanding the rhetorical situation through the lens of its affect on the body. Remember that Blair's analytical framework for material rhetoric is anchored in five main questions that consider the rhetorical text from the vantage point of its embodied, contextualized interaction with the world. For example, she asks: "(1) What is the significance of the text's material existence? (2) What are the apparatuses and degrees of durability of displayed by the text? (3) What are the text's modes or possibilities of reproduction or preservation? (4) What does the text do to (or with, or against) other texts? (5) How does the text act on people?" (30). In this chapter, I build upon and extend this framework in such a way that also accounts for the impact of the multimodal text on the technologically-mediated, posthuman body. In this context, for a visual-material rhetoric to remain mindful of embodied experience, it must avoid the pitfalls of conflating posthumanism with "transhumanism (the biotechnological enhancement of human beings) and narrow definitions of the posthuman as the hoped-for transcendence of materiality" ("What Is Posthumanism?"). Subsequently, the GPS, when understood as contributing to posthuman, embodied ways of knowing, would not constitute a kind of prosthetic extension of the body that transcends materiality, but rather it would serve as a rhetorical artifact that engages the body. Through its multimodal qualities, such as its physical presence and its use of visual and audio cues, the device engages users in a unique way that elicits interaction and bodily engagement in everyday settings and mundane activities. In demonstrating the implications of visual-material rhetorics for the posthuman, mediated body, I do not attempt to rehearse the trajectories of theories of the posthuman or re-conceptualize these concepts, per se. Rather, I begin by exploring a useful question posed by John Muckelbauer and Debra Hawhee, who, in contextualizing their inquiry, first ask their readers to consider "posthumanism as an attempt to engage humans as distributed processes rather than as discrete entities. In doing so [. . .] posthumanism 'emerge[s]' at nodes where bodies, bodies of discourse, and discourses of bodies intersect'" (768). These intersections, write Muckelbauer and Hawhee, complicate traditional notions of audience, purpose, and context. Thus, they ask: "With the emergence of posthumanism—which challenges distinctions between subjectivities and consequently renders the notion of persuasion unclear—what becomes of rhetoric?" (770). In response, I suggest that visual-material rhetorics can provide a helpful point of entry into such questions by examining the implications of multimodal rhetorical artifacts for contextualized, embodied experience and the relationships that are forged through such experiences. Implicitly aligned with this idea, Collin Brooke notes that "posthuman rhetoric, as a return to embodied information, involves a revaluing of partiality. A posthuman rhetoric would allow us to turn our backs on omniscience and the humanist values of mastery and control that derive from the will to knowledge" (791). In a discussion that also seems to resonate for Brooke (781), Hayles describes her vision for or definition of the posthuman when she writes:

> If my nightmare is a culture inhabited by posthumans who regard their bodies as fashion accessories rather than the ground of being, my dream is a version of the posthuman that embraces the possibilities of information technologies without being seduced by fantasies of unlimited power and disembodied immortality, that recognizes and celebrates finitude as a condition of human being, and that understands human life is embedded in a material world of great complexity, one on which we depend for our continued survival. (*Posthuman* 5)

When humans use the GPS to mediate their personal navigation, they arguably engage in acts of resistance, negotiation, and embodied experience that at once "embraces the possibilities of information technologies" while implicitly remaining mindful of the complex material world to which Hayles refers. These engagements with the GPS and with their physical surroundings constitute acts of purposeful decision-making that, again, for Kim et al., are also related to the idea of agency. That is, in a recent study, Kim et al. describe how multimedia tools can "move the user as close as possible to informed decision making" and "support the autonomy" of the users of a technology (341). Kim et al. define autonomy as referring to "an individual's right to self-government" (341). While their focus is primarily on the idea of informed consent and the communication that takes place with patients and parents in healthcare settings, they are helpful in noting that "to exercise autonomy," users "must engage in agency"; they then define agency as a person's "ability to recognize and understand options, recognize resources that facilitate decision-making as well as barriers to informed choice, and make purposeful decisions" (341–342). While the GPS may play a role in moving users to engage in agency, I suggest more explicitly that the GPS participates in the co-construction of agency through an "interactive process" involving multiple levels of exchange (Kennedy 308). As a visual-material rhetorical approach helps show, the GPS summons the user and invites an audience—in some cases even a dialogue—that requires the user to remain mindful of both symbolic and material environments. The user's simultaneous interaction with the symbolic, virtual displays of the GPS and the physical terrain of the material world constitutes a multimodal, interactive engagement that participates in the construction of geographic knowledge that informs users' decision-making. In this way, as Karlyn Kohrs Campbell describes, agency is "constituted and constrained by externals that are material and symbolic," and as Kennedy also points to in Campbell's work, agency is "protean, ambiguous, open to reversal" (Campbell 2). Along a similar register, Carolyn Miller describes agency as "the *kinetic* energy of rhetorical performance," and as "positioned exactly between the agent's capacity and the effect on an audience" (147). This idea is compatible with an understanding of the GPS as facilitating purposeful decision-making and fostering or even constraining geographic knowledge through interactive texts that summon or invite an audience, and are malleable, open to negotiation, and able to be subverted or resisted by that audience. The idea of the GPS and its user as participating in the construction of agency is also compatible with the more nuanced sort of embodiment called for by Hayles in her discussion of the posthuman body.

Five years after the publication of *How We Became Posthuman,* Hayles acknowledges in *My Mother Was a Computer* that contemporary debates about the posthuman are likely to focus "on different versions of the posthuman as they continue to evolve in conjunction with intelligent machines" (2). While Hayles maintains her earlier commitment to the "importance of embodiment," she rightly cautions against a binary reading of embodiment and disembodiment, instead describing this distinction as having "fractured into more complex and varied formations. As a result, a binary view that juxtaposes disembodied information with an embodied human lifeworld is no longer sufficient to account for these complexities, [and] contemporary conditions call increasingly for understandings that go beyond a binary view to more nuanced analyses" (2). Hayles goes on to note that such an understanding "requires repositioning materiality as distinct from physicality and re-visioning the material basis for hybrid texts and subjectivities" (2). The GPS illustrates this point well, for while it is indeed a physical object in itself, it is not defined solely in terms of its physicality; in other words, its physicality does not constitute the whole of its material impact on the body or its relationship to the user or to knowledge-making. Instead, its physicality is but one component of its functioning as a multimodal artifact of visual-material rhetoric.

In this study of the GPS as multimodal rhetorical artifact, then, I asked the following questions: 1) In what ways do the various cues provided by the GPS resonate with and influence its users? 2) How do GPS users engage with the GPS? 3) What are the potential consequences of the GPS on geographical understanding? 4) What are the broader consequences of the GPS on the body and for purposeful decision-making? 5) How does understanding the GPS as a visual-material rhetorical artifact help demonstrate the value of visual-material rhetorics as a project of inquiry that can function across contexts and situations?

ON CHAPTER ORGANIZATION AND DATA COLLECTION

To address these questions, I designed a qualitative study and conducted face-to-face interviews of approximately 30 minutes with twenty-two GPS users.[2] I conducted my research at a small liberal arts college in the mid-Atlantic region, and delimited the parameters of my sample to any faculty and staff at the college, including part-time adjunct faculty and lecturers, as well as the family members of any faculty or staff who work there.[3] In my sample, 59% of participants were female and 41% were male [4]; 95% were Caucasian and 5% were African American; 43% of participants were adjunct faculty, 39% were college staff, 9% were full-time faculty, and 9% were family members of faculty or staff; 59% of participants owned a GPS, 36% borrowed a GPS from the library, and 5% borrowed a GPS from a family member who owned one. Of the participants who owned a GPS, 31% had owned a GPS for three years or more; 31% had owned a GPS for two years or more; 30% owned a GPS for one year or more; and 8% had owned a GPS for less than one year.

Participants who volunteered to be in the study were asked to use an in-car GPS navigation device for a two-week period during any driving they did. If a participant owned a GPS, they were free to use that one. If a participant did not own a GPS, they were able to borrow one through the GPS borrowing program at the college library.[5] After they had completed the usage period, I interviewed each participant about their experience using the GPS.[6] (See Appendix A for a list of my interview questions.) I tape recorded and transcribed each interview verbatim immediately after the interview. After all interviews were completed and transcribed, I coded the transcriptions and looked for themes that emerged. [7] (See Appendix B for a table containing my coding categories.)

It is important to note that it is not my goal within this chapter to conduct a usability study of the GPS for new and experienced users or to conduct any sort of feature comparison across GPS brands. [8] Rather, my goal has been to capture a diverse range of GPS experiences across users, and to see what visual-material rhetorics can reveal about the impact of these technological contexts on the mediated, posthuman body. This is not to say, however, that the experiences of newer and more seasoned users are identical, or that differences therein should not be acknowledged. In some cases, these categorical distinctions seem to account for different types of bodily experience or influence of the technology on geographic understanding. In such cases, I am mindful of these differences and acknowledge them specifically within the context of their value to a visual-material rhetoric approach. In fact, as I coded the interview transcripts and looked for the themes that emerged, I was attentive to these potential differences and created the participant key such that the comments of newer users could be distinguished from those of more experienced users and invoked if necessary. [9] I have concealed, however, any specifically identifying information about participants, both in terms of personal names or references, and in terms of any specific information related to addresses or particular place-names described. I refer to names of towns, cities, restaurants, stores and other common or generally known sites only when such references would not reveal any specifically identifying information.

This chapter again employs a Blairian-Foucauldian interpretive framework, this time for understanding the consequences of in-car GPS systems on the mediated, posthuman body and for visual-material rhetorics more broadly. I let the major themes that emerged during the coding of the interviews, along with the common trajectory of participants' descriptions of their experiences with the GPS, drive my organization of the chapter. Thus, the chapter generally moves from a discussion of the physicality of the GPS, to the impacts of its visual and audio cues, to the broader implications for how the GPS and the user participate in purposeful decision- making and geographical understanding.

As a heterotopic artifact, the GPS produces digital, cartographic texts that at once represent or bear relation to the surrounding environment while also challenging other texts or geographic modes of knowing. Likewise, as mentioned earlier, Foucault suggests that we reside in heterogeneous spaces that, while perhaps sharing common environmental or spatial characteristics, "are irreducible to one another and absolutely not superimposable on one another" ("Of Other Spaces" 23). He calls these sites heterotopias, and says that while they may be in relation with the territories that they represent, they may also "suspect, neutralize, or invert the set of relations that they happen to designate, mirror, or reflect" (24). I argue that the GPS map, like other maps discussed in this book, functions heterotopically to both identify and complicate the spaces being characterized and the internal relationships that constitute their rhetoricality and sense of place at an individual or personal level. In other words, the map produced by a GPS may mean different things for different users, each of whom will bring their own knowledge, judgment, and expectations to bear on it. The map may represent an identifiable and expected physical territory, but depending on the user's preferences and processes for working with the map, it may come to reflect different ideas and may be used in different ways.

As an artifact of visual-material rhetoric, the cartographic texts produced by the GPS may be viewed as durable, reproducible to a certain extent, and may supplement or be supplemented with other types of geographic knowledge claims. Additionally, Blair notes the material text's potential for "enabling, appropriating, contextualizing, supplementing, correcting, challenging, competing, or silencing" (39). The GPS engages the user, and in doing so, enables and invites an audience, even a dialogue. Some users tend to anthropomorphize the GPS or forge what appears to be a personal relationship with the artifact. Users' interactions and communication with the device also provide a means for challenging or correcting it, in order to assert their own claims to knowledge and engage in decision-making. In this way, the map would seem to open up the possibility for resistance, inviting users to identify "the nonlegible practices that are performed within the weave but are asymmetrical to it" (Biesecker, "Michel Foucault" 357). The map, for instance, is indeed a representation of a particular, physical territory and often conveys multiple ideas about a place. These representations of place are produced through power relations, and thus may be understood as visual-material heterotopic texts.

The GPS directs the user's attention in ways that are sometimes helpful and sometimes potentially distracting. Nonetheless, all of these features allow the user to carry out actions that have consequences; that is, some users express that, with the GPS, they are no longer afraid of getting lost, while others express that they will not give up their paper-based maps or computer-generated directions. In this way, the GPS may both enable and constrain understandings of the surrounding environment or a willingness toward exploration. The GPS has varied impacts on the user's comfort or anxiety levels related to driving and travel, and subsequently impacts social relationships, interpersonal relationships, and everyday understandings about and decision-making within the world. With these issues in mind, this chapter examines GPS use and its function in everyday life, subsequently demonstrating the value of visual-material rhetorics in helping to illuminate the everyday contexts influencing the technologically-mediated, posthuman body.

The Physicality of the GPS

In describing their experience with the GPS, many users were quick to note the physicality of the device and the ways in which the GPS, as a tangible artifact, would impact their driving routine. One user described the ebb and flow of interest in the device, and the concerted effort it sometimes takes to remember to use it: "We got ours for Christmas last year, so we started using it then. [. . .] My husband was really into it at first. [. . .] I used to leave it up on our dashboard, and then I put it away, so I don't always take it out. [. . .] sometimes it's more of an effort to use it" (OA5). In this case, the user's interest in the device manifests in different levels of physical interaction with it; these physical interactions involve the conscious placement of the GPS on the dashboard, and likewise the conscious removal of the device from its designated place in the car. Once the device is put away, it becomes an effort to put it back. Another user described the system for keeping and using the GPS in the car: "I just plug it into the cigarette lighter [. . .] the armrest flips up and there's a little storage thing inside, so I just drop [the GPS] back in, and then the mount goes under the seat"

(OSA11). Here, the user's description of "just" plugging the GPS into the cigarette lighter, and "just" dropping it into the armrest implies a simplicity and ease of use that makes the act of handling the GPS seem like an appealing part of the driver's daily routine. One user described being so accustomed to seeing the GPS that it now feels as though something is missing if they are in a car without one: "[I]f we get into somebody else's car and they don't have it, it just seems like something's really missing" (OFT2). Similarly, another user commented: "Most of my travels in general are within [the city], so I'm pretty familiar with where everything is, but I tend to always still have the GPS on, it's almost like a reflex" (OSA6). A newer user described having "trouble finding a place to put it," and "didn't want to use the kind that, you know, the mount that sticks to my dash," but also noted that as the GPS "became more second nature, it was less of a distraction, so [. . .] it became more part of my daily [routine], part of the car" (BA14).

These accounts help demonstrate the ways in which the physical tangibility of the GPS impacts drivers' routines, even prior to turning on the device and navigating with it. Michael Knievel feels that acknowledgements such as these are often under-acknowledged within fields like rhetoric and technical communication, and sees a need to better account for the materiality of technology in his debate about whether technical communication may be understood as humanistic or instrumental discourse. Knievel describes "a continuum of technology definitions" that perceive, on the one hand, a "view of technology as a material artifact—an instrumental interpretation in its simplest form—and on the other, that of technology as a textual entity or rhetorical, cultural force—an instrumental ethos governing social systems, economies, and so forth" (70). He feels that definitions which understand technology as discursive are familiar enough within the humanities, but those that view "technology as a physical artifact" are less often embraced (71). If discussions of technology's materiality are left tangential or made too implicit, Knievel questions whether fields like technical communication "can legitimately claim humanistic status at all" (72). In linking Knievel's ideas even more closely to a discussion of visual-material rhetoric, I suggest it becomes helpful to also understand "materiality as distinct from physicality," or physicality as but one component of materiality (Hayles, *My Mother* 2). That is, the GPS's physicality does not account for the whole of its material impact. Users describe having to contend with the artifact's physicality (figuring out where to place it in the car, remembering to put it there in the first place, plugging it into the cigarette lighter, and stowing it in the armrest) even prior to describing its broader implications for decision-making and the practices that inform embodied knowledge. And since part of the act of reading the multimodal GPS text involves the need to see the device as well as hear it, then its physical accessibility and the user's ability to work with its physicality becomes just as meaningful as the interactive, multimodal cues that contribute to rhetorical understanding. To understand artifacts like the GPS as simultaneously physical and visually and materially rhetorical not only helps to abate concerns about the perceived subjugation of "the idea of technology as a material artifact" within fields like technical communication (Knievel 72), but also provides a productive approach for analyzing multimodal, visual-material texts within the disaggregated settings that Warnick and Hayles describe.

VISUAL CUES AND THE BODY

The GPS is a visual-material rhetorical artifact that produces multimodal cartographic texts. Consistent with Turnbull's understanding of the map, these texts produce "a graphic representation of the milieu, containing both pictorial (or iconic) and non-pictorial elements," and "directly represent at least *some* aspects of the landscape" (3). Lisa Parks also describes the GPS as a device whose interactivity is rooted in the relationship between meditated representation and bodily movement:

> What is significant about GPS technology is that it both locates the user at an interface and presents a mode of interactivity that is predicated on the body's movement rather than its motility. If we accept the definition of an interactive medium as 'one in which the user can influence the form and/or content of the mediated presentation or experience,' then GPS images certainly qualify. For the user modifies the visual display of the GPS receiver each time she moves through space. In this case, then, satellite interactivity involves the practice of using a GPS receiver to plot one's own movements and to generate a visible register of that movement. (211)

The visual texts displayed by the GPS are indeed an integral component of both its multimodality and its impact on the user's bodily experience. In fact, many of the GPS users interviewed in this chapter were quick to describe visual aspects of the GPS that resonated with them in one way or another.

Visual Displays as Providing Validation or a Sense of the Surrounding Landscape

Many users described the general aesthetic appeal of the visual display or the way it would allow them to see the surrounding landscape in relation to where they were located. One user described the visual display as comforting:

> [C]uriously enough, even when I'm going somewhere where I'm completely familiar, like the supermarket or the gym or something like that, I don't even get past our street without pushing on the button that brings it on, because I don't like seeing the empty screen. [. . .] But I would say that when I have it on, when I'm going to a familiar place, I don't really need to have it on—I just like it. (OFT2)

Another user described a similar notion: "I liked just the map function and being able to see where I was in relation to the surrounding countryside" (BA14). The idea that the GPS's visual display provides the user with a sense of their location relative to the surrounding landscape is similar to the ability of wayfinding signs, as described in chapter three, to foster cognitive or mental maps that aid "in the development of spatial understanding and spatial ability" (Xia et al. 446). Additionally, one user described the sense of validation provided by the GPS's acknowledgement of changes in the landscape and the comfort of seeing their trajectory displayed on the screen: "My husband I think pays more attention to the screen [. . .] ours changes when it gets darks out, so he notices that change, or if we go over a river, you know, it shows the river, so he pays more attention to that. I mean sometimes it's nice just to watch it [. . .] to view the map and see where I'm going" (OA5). In these instances, users understand the symbolic cues provided by the screen as providing information about their physical surroundings that implicitly furthers their geographic knowledge. This information is not only viewed as interesting on a more esoteric level but may also become comforting or validating for the driver as it allows for the development of cognitive maps of their surroundings. These visual cues then not only contribute to the interactive nature of the GPS but also provide information that may ultimately play a role in or facilitate navigational decision-making.

Visual Displays as Providing Navigational Cues

Several users noted that the GPS allowed them to see more specifically where they would need to turn. One user, for example, described being able to visualize where to turn based on what the road looked like up ahead:

> I'm very visual, in the way that I deal with things, or even in the way that I get around, so to be able to, you know, in my head, if I'm driving on a road, somehow I always kind of picture it as straight, but to see on the GPS, even the curves, I really actually liked a lot. Not that it changed the way I drove drastically, but as you're getting closer to things, you could see, 'oh, well, it's after that curve, that place where the road curves to the left,' you need to start carefully watching for the turn—things like that, so I liked that a lot. (BSA15)

Another user noted that the visual display made them less likely to miss a turn:

> [When] I'm invited somewhere, where I'm not familiar with that road, I really like seeing [the display], because if I'm headed toward a part of [the city] that I'm not that familiar with [. . .] then I can actually see the specific street coming up and I know immediately, oh, that's where I have to turn. And then I don't have to be concerned about, oh, did I already pass it? I'm not sure if I passed it, or somebody gives me directions, but then I think, well, was that what they counted as a traffic light, or was that really not a traffic light? Or how many stop signs, instead of having to worry about counting stop signs, I can just be, kind of watching for the road. So that's how I've used it around town. (OFT2)

One user not only described the GPS as allowing them to better learn landmarks in the physical landscape but also described the GPS's highlighted route as easier to follow than markers in the physical landscape:

> I also like [. . .] being able to look ahead on partly familiar routes, where I can say, 'the turn that I need is not this one coming up, it's the next street,' so that I can learn landmarks. And it's much easier to say 'the little red route goes over there,' than trying to look at the tiny little signs, especially at night (OA1).

Again, these users describe the visual cues provided by the GPS as aiding in informed decision-making relative to their navigational choices. Agency, in this case, may be understood as the sort of interactive process that Kennedy describes, and as constituted not only by the symbolic cues provided by the GPS but also by the user's understanding of those cues and how they pertain to their physical surroundings.

Users also implicitly commented on the interplay of visual and textual information in the GPS display: "I do like how you can actually look at it and it'll say, 'nine miles until exit 22'—that I really like—that's helpful for me" (OA3). In drawing the user's attention to specific geographic information, then, as Blair has put it, the GPS "directs the vision to particular features" or information about the landscape (46). In directing vision toward specific features of the landscape, however, the GPS also employs selectivity, excluding and including aspects of the landscape through the use of graphical features to attract drivers' attention.

Graphical Features, Selectivity, and the Symbolic Environment of the Visual Display

One user, for instance, described the use of the color green and cultural symbols such as a cap and gown as aiding in their ability to make inferences about the sort of community through which they are traveling:

> I like knowing when I'm going through a new place, oh, this community seems to have a lot of green space, because you see it—it shows up as green, or, oh, this place must have a college, because you see a little cap and gown thing in the middle of a big space. So I like knowing, it gives me more of a sense of what the community must be like because it has a lot of parks or because it doesn't have any, or, or it has a river running through it, you know, it doesn't mark every single thing but it marks golf courses, it marks colleges, [. . .] it just makes me

feel like I understand what's around me better than if I'm just on a highway, focused on the road that I'm on. (OFT2)

Another GPS user commented on the use of color as both helpful in describing the environment and in its impact on their bodily experience: "I do like the fact that when the sun goes down it's got a feature where it will automatically change the colors, so there's an automatic night display, so it's not going to ruin your night vision quite so much" (OA1). One user also described getting lost in terms of how their car was represented symbolically through the visual display. For example, as this user described, they were driving on back roads when the GPS did not produce maps specific to their location: "I was on the road, but there was no map to tell me where I was. [. . .] I was on white—I was no longer on a road—there was no gray or pink line to direct me" (BA21). Eventually, this user had to stop and ask for directions before continuing home. While all of these users, to some extent, describe the physical experience of driving relative to their interaction with the symbolic features of the visual display, this last user's description of getting lost as driving "on white" bears a particularly close analogue to that of someone describing a gaming or virtual environment.[10] Kevin Moberly, for example, understands video games as "evaluat[ing] players on their ability to compose themselves in relationship to these highly symbolic environments—to write and ultimately revise their actions in relationship to the reality that is manufactured on the screen" (291). The GPS, too, requires that users compose themselves in relationship to highly symbolic, selectively represented routes, locations, and place names. These symbolic representations then have clearly material impacts, which may result, as it did in this user's case, in having to supplement the cues of the GPS by stopping to ask for directions.

Another user described experimenting with different ways of programming the GPS to include or exclude the visual display, and their subsequent preference for the guidance provided by the colored line indicating the route:

> I have tried using just the step-by-step [directions], or where it just shows the arrows, and you're going to turn in 7.5 miles at this intersection, where it's just the words and the directions, like you would get on a MapQuest, versus the actual map where it shows the little vehicle and the pink highlighted route—I think

> I like the visual, with the pink highlighted route. And the audio comes with that, but I do prefer that. (OA7)

The highlighted route functions as a prescribed pathway that directs users to their destination, and as this user notes, the visual display of the highlighted route does not function in isolation from other cues. It is accompanied by textual and audio cues that, together, create a selective, symbolic representation of the surrounding environment that influences self-navigation. These selective cartographic representations can be both helpful and distracting depending upon the contexts influencing the user's experience.

The Interplay of Visual and Audio Cues and Possibilities for Distraction

Users may pay more or less attention to particular aspects of visual and audio cues depending upon their individual contexts and how comfortable they are with their surroundings. One user, for example, described attention to visual and audio cues as dependent upon when a turn was coming up: "I think the audio cues help me to stay in tune with where I'm going. [. . .] But I think in [. . .] shorter areas, I usually pay closer attention to the arrow a bit more, because it can tell you, 'turn right,' but actually that's in 100 yards, not right now. So I do tend to watch the screen a bit more in closer, tighter spaces" (OSA13). Another user described paying more attention to the visual display when the car's radio was on: "[S]ometimes if we had the radio on, you couldn't hear the audio, so then we were just looking at the video" (BA22). This description of the GPS audio as competing with the car radio raises the issue of the distractions potentially posed by the devices. Such distractions tended to be articulated more frequently by newer users.

Understandably, newer users tended to comment that the cues provided by the GPS could be distracting at first, and in this way, the user's prior experience with the GPS certainly plays a role in its potential consequences on the body. One new user expressed concern that the GPS could be used while driving in the first place, and said: "I probably had almost had two accidents the first day I had it. Just because I was looking at *it* as much as I was looking at the road" (BA14).[11] This user also commented that the physicality of the device itself initially impeded visibility: "I'm also nearsighted [. . .] so it was hard to figure out where to have [the GPS] so I could see distance and still not have to crank my head to see the small picture close up. So it was difficult finding a place to locate it, where it was usable and not in the way" (BA14). A more experienced user commented that they had to be careful not to "put more attention on the GPS than on the driving" (OSA9). Another user recalled their fascination with the GPS when first using it: "Yeah, in the beginning, [. . .] because I wasn't used to it, I think I may have been a little too distracted by it. I don't remember ever even coming close to being in an accident but I would become aware of, 'oh, I'm probably looking at this too much,' because it was fascinating" (OFT2). Working in the negative, then, the GPS may serve to interrupt the user's navigational experience by creating visual, auditory, or other physical distractions; working in the positive, it may interrupt the navigational experience by introducing new or engaging information about the driver's physical surroundings. The GPS then not only directs the driver's attention to specific aspects of the surrounding landscape through its selectivity but it also competes for attention with other contexts influencing the driver's experience, such as their relative newness to the technology, level of curiosity about the surrounding landscape, or even other competing modalities such as the volume of the car stereo.

Audio Cues and the Body

Users' responses to the audio cues provided by the GPS were often linked to their knowledge of the surrounding area. That is, many users tended to appreciate or pay greater attention to audio cues when they were less familiar with their surroundings.

Audio Cues as Both Helpful and Bothersome

As one GPS user described: "[W]hen I knew where I was going, [or] I knew I had to turn left on a certain road, and it would say 'turn left in 800 feet' or something, that was distracting—I knew that. You know, so the voice coming at me, telling me, was just sort of something I didn't need to know" (BA16). Likewise, another user said: "Sometimes I'll turn it off so it's not talking to me if I know where I'm going" (OA3). In an "ethnographically-informed" study of ten GPS users, Leshed et al. also note that informants reported finding the "directions 'annoying' in familiar neighborhoods or on known routes and chose to turn the voice off" (1, 5). Users did comment, however, that they appreciated verbal reminders when, say, an exit was approaching: "So you try to limit the voice cues, but obviously if it's really busy, it kind of does help to kind of have a reminder that you have an exit coming up or something" (OSA6). In limiting voice cues or muting the GPS, this user engages in a sort of negotiation with the device by deciding when and to what extent they want to engage with its interactive information. One user described the verbal reminders of an approaching exit as providing a sense of safety while driving in bad weather conditions: "I remember once having to go out to a gallery, and it was pouring, pouring rain. And the rain was coming down so heavy, there was no way I would've caught some of the streets and the exits if it wasn't for the GPS saying, you know, 'in 0.8 miles, turn right'" (OA7). In these ways, then, the audio cues provided by the GPS serve as catalyst for navigational decision-making and may detract from or enhance the navigational experience depending on the user's individual context.

Audio Cues as Summoning the User, Inviting an Audience, and Fostering Relationships

Perhaps the most interesting consequence of the audio cues is the way in which they summon the driver, inviting an audience, and in some cases, even a dialogue, with the GPS. Users are often aware of this aspect, acknowledging both the engaging and sometimes humorous nature of the audio cues.[12] One user noted the challenge of finding an appropriate voice setting: "Well, we did find that we liked certain accents better than others. And I know that sounds ridiculous, but the voice I initially started with did make me feel like, 'Ah, you're stressing me out!' But then we found another voice that was much more appropriate. So we did like the feature of being able to change that" (BSA17). Another user mentioned that "you find yourself kind of talking to the unit eventually" (OSA6). Interestingly, one user explicitly described the ability to choose a voice setting as tied to the tendency to develop a relationship with the GPS: "[A]fter using it long enough, the GPS has become a person. And I've kind of noticed, that myself, it's almost like it's interacting with the person even though it's electronic. But I think the fact that you can choose your voice and it's a very soothing voice, you kind of develop that relationship with the machine" (OSA11). This user also commented on the choice of voice as implicitly contributing to the experience of using the GPS: "Actually, I have a very calm female voice on. And it's one of those things, it's very laid back (OSA11). While it is beyond the scope of this chapter to engage in an extended analysis of the implications of the gendered or culturally diverse GPS voice settings from which users may choose, it is clear that users tend to choose a voice that resonates with them or makes them feel comfortable.

Choices related to gender and voice settings then impact the user's ability to interact with the GPS, which sometimes leads to an anthropomorphizing of the device, or a projection "of a human behaviour onto a nonhuman [. . .] object" (Johnson [Latour] 265).[13] Latour sees anthropomorphism as underscoring an "untenable" divide between humans and nonhuman technical objects (266). That is, on the one hand, extrapolating from his well-known discussion of the sociology of the door-closer, the GPS may be understood as already anthropomorphic, in that it has been invented by humans and shapes human action (265–266). On the other hand, Latour also understands the door-closer as "substitut[ing] for the actions of people, and [as in the case of the door-closer] a delegate that permanently occupies the position of a human" (266). The GPS does not stand in for human action in quite this way—while the GPS, as a multimodal technology, certainly mediates the actions of its users and thus serves to muddy the divide, it does not substitute for human action, permanently occupy the position of human, or take on full agent status. Instead, users more typically describe an ongoing set of interactions, negotiations, and exchanges with the device that serve to co-construct agency and influence geographic knowledge-making. Visual-material rhetorics can then help reveal how these interactions with the GPS may be implicated in the co-construction of agency and understood as a response to how the GPS, as a visual-material artifact that engages the mind and body, summons the user or invites an audience by way of its visual and audio cues, ultimately impacting bodily experience. Moreover, the articulation of a sort of relationship with the GPS provides users with a means for pushing up against or resisting the knowledge claims implicit in or inferred through its multimodal cues and directions.

WORKING WITH AND AGAINST THE GPS

Understood through a Blairian lens, the GPS functions rhetorically by challenging and sometimes correcting users' knowledge of their environment. It invites an audience not only through its multimodal cues but through the directions that it implicitly asks users to follow. In this way, the GPS directs the body and prescribes a pathway or a destination. To follow these directions is often perceived as a high-stakes endeavor, especially when users feel that they know a more feasible route or one that would better serve their individual needs. Users will then engage with the GPS in a variety of ways, many of which involve some form of a dialogue, argument, or seemingly personal interaction or negotiation with the device. Again, the expression of these verbal interactions reflects the broader impact of the GPS on the body and its work as an object of visual-material rhetoric that has the ability to inform or facilitate navigational decision-making.

Resisting the GPS

Many users, for example, expressed a sort of power struggle with the GPS or a hesitance to "listen to" the device and the ways in which they would challenge its claims to knowledge. As one user put it, "I do still check to make sure that it's giving me good information. [. . .] I'll go and say, 'No, I'm not listening to you,' as far as that" (OA1). Another user described a more explicitly adversarial interaction: "[W]e decided to go a different way that we knew, and the GPS was like fighting us, you know, for maybe six or seven miles" (BSA20). When reflecting on and assessing the cumulative use of the GPS over time, one user commented: "I think she and I are still fighting it out" (OSA8). Another user described watching her parents' adoption of the GPS and the ways in which her father would challenge the device: "And it was funny, because when they first got it to use with their camper, and they were testing it out with the car and driving with the camper, in the beginning, my dad argued with the GPS all the time. [. . .] [H]e always thought he wanted to take a different route [. . .] because he had spent a lot of time on the road as a salesman" (OSA11). Users were frequently very expressive and emphatic in describing these power struggles, largely because the decision-making that often accompanies these debates has a very real impact on users' physical experience in the world. Interactions with the GPS, then, are not taken lightly and are often described in terms of arguments with or resistance to it. In this way, the GPS text and its multimodal cues function as an interactive heterotopic space, in which particular geographic spaces are represented and contested. These contestations open up spaces for resistance that may result in the broadening or constraining of geographic knowledge, or play a role in decision-making that ultimately has an impact on the mediated body.

"Recalculating, Recalculating"

The GPS often engages in a variation of what Blair refers to as the act of "correcting," or in GPS parlance, the action more commonly referred to as *recalculating* (39). The recalculating function understandably elicits different reactions from different users; nonetheless, it was common for users to describe an interaction with the GPS that involved the act of recalculating and its impact on both the mind and body. One user who felt familiar with the surrounding area and did not want to take a different route described active resistance to the GPS: "The one time I used it for a long distance thing where I kind of knew where I was going, it was telling me to go a way that I wasn't at all familiar with, and I didn't—I didn't do it—it kept recalculating, because I kept going the way I thought it was" (BA16). Miller also acknowledges "resistance models of agency, models that usually rely on a metonymy between agent and agency" and are produced through attempts to resist perceived societal pressures or "structures of institutional, corporate, and ideological domination" (144). This user's choice to continue on the route they thought was right, thereby ignoring the GPS's attempts to recalculate, constitutes an act of resistance. In this case, the user exercises informed decision-making that is enabled through an interactive sort of agency that is constituted through and within the exchange between the GPS's directive and the user's resistance to it. Agency may then be understood both as an act of resistance and as happening in the "kinetic energy of rhetorical performance" (Miller 147).

Another user interpreted the rerouting feature as adapting to their needs and described feeling empowered by it: "I love how it suggests an alternate route. I loved how it would recalculate and adapt to me—I was in control" (BA21). This user also implicitly understands geographic knowledge-making as an interactive process that involves a negotiation with the GPS. By attributing the GPS with the communicative ability to "suggest" a route, this user also describes the GPS as working "in the service of agency" (Kennedy 305). That is, the GPS is obviously not "sentient in the ways that we normally think of agents," but may be understood as "perceiv[ing] their environment and initiat[ing] action with it" (Kennedy 306). Thus, the GPS may be understood as co-constructing agency and enabling change in terms of the navigational decision-making that it facilitates. Another user engages in a sort of self-deprecating banter with the GPS and admits to feeling at fault when being rerouted: "You make a mistake, it tells you, 'recalculating route,' which is a way of saying, 'idiot, you screwed up,' or if you really blow it, we've noticed, 'make the first legal u-turn'" (OSA4). Here, this user more explicitly anthropomorphizes the GPS, thereby ascribing it with a human capacity for dialogue that further reinforces its agent status and ability to participate in decision-making by way of correcting that ultimately impacts the body.

In each of the examples described here, users' responses to being "corrected" by the GPS involve some form of interaction with the device and subsequent decision-making that has consequences on the body. Users may feel put off or like they are being told what to do and choose to continue on their path despite the directive; they may accept the cue as a suggestion and feel more in control of their experience, or they may be rather dismissive of the GPS and treat it as a good-natured passenger providing helpful guidance. In any case, the GPS and its user co-construct an interactive agency that is malleable and involves a process of ongoing exchange, negotiation, and sometimes resistance. The GPS then has the capacity to engage the user with its cues and directives that may either facilitate or discourage decision-making. Understood from the vantage point of visual-material rhetorics, these decisions ultimately bear consequence on the body. Moreover, it becomes possible to see how users resist the GPS through embodied practice that, as Hayles has put it, "moves in conjunction" with technology and is mindful of "discursive constructions" (*Posthuman* 195).

Subverting or Challenging the GPS

Many users also negotiate with the GPS through processes similar to what Blair has called the act of "challenging" (42). These GPS users, who are often skeptical of the accuracy of its directions, describe working within the perceived limitations of the device in order to challenge and subvert its cues and achieve a desired result. One user, for example, learned how to manipulate the GPS such that its imperfections became opportunities for positive decision-making: "You know, every now and then it does go wrong, but usually what'll happen is, if it doesn't work, you just, 'okay, well I can't turn there because there's not a road,' so I'll just go up here. It'll say, 'recalculating,' and it'll work you back around in a way that works" (OSA11).

Another user described using a combination of local knowledge and the GPS's directions to navigate to a destination. This subversion entailed the more authoritative position of ignoring the GPS until it adapted to the driver's path:

> For instance, going to the beach, which was [in] New Jersey [. . .] there's kind of two routes that we've historically taken, and one's down through Red Lion and Port Deposit, to catch one of the major highways near the Delaware Bridge, but this time I decided we were just going to take 83 down, and you know at first that contradicts the instructions on the GPS; finally, the GPS gives up and plots it the way I want to go, once I get far enough that it doesn't make sense to go the other way to the GPS. And so we followed that, we did that on the way down to the beach. (OSA6)

Again, to describe the GPS as giving up or surrendering to the user affords it an ability to engage with the driver in a negotiation about the route to be taken. One user explicitly described learning how to manipulate the GPS in order to change routes:

> I thought I knew where I was going, [and] couldn't go any further because there was a marathon going through, the road was closed, then I said, 'oh heck, I need to get to such and such.' You know, it recalculates, so if I would turn on to another road, it would give me an alternate route, so I kind of learned how to, even though I couldn't say—I didn't know how to say—'Hey, dude, I need an alternate route,' I found that if I just went on a different road, of course it gives you an alternate route. So I learned how to kind of manipulate it a little bit. (BSA15)

This driver too describes a dialogic negotiation with the GPS that involves circumventing its symbolic constraints by making a navigational decision in the physical world; this navigational decision then has both corporeal impact and allows for the continuation of the interactive exchange between the GPS and the driver. And as another user described, "[S]ometimes, I want to take a scenic route, I have to do a little research up front, to know which way I want to go so that the GPS doesn't take me *its* way. But again, if you know that up front, you just work with it" (OSA9). The negotiation described by this user also entails supplementing the directions provided by the GPS with personal knowledge of or research about the surrounding area. And indeed, the act of supplementing speaks to the rhetorical work of the knowledge claims implicit in the cartographic text of the GPS and its subsequent corporeal impact.

Supplementing, Resisting, and Competing with the GPS

Understood through a Blairian lens, the GPS also functions rhetorically by supplementing users' own local knowledge as well as the directions provided by paper maps and online programs such as Google Maps or MapQuest. Conversely, users may also supplement the directions provided by the GPS with their own knowledge or research to create individualized routes. As Blair notes, the act of supplementing turns the "completed text" into an opportunity for contextualized, individual practice (40). As many GPS users described, the act of supplementing typically entails an individualized negotiation with the GPS. This individualized negotiation often begins with a general decision on the part of the user. For example, a GPS user might first make the decision to use the device only in particular circumstances, such as on longer trips. During this time, they may work with or against the GPS, thereby engaging with, negotiating, or resisting the knowledge claims implicit in its multimodal cartographic texts.

When navigating with the GPS, many users noted that they were more likely to use it on longer trips, when they were going to a new or unfamiliar place, or to find restaurants or other points of interest while they were on the road. One user remarked, for example: "I've been using it on and off mostly for some longer trips that I'm not really sure where to get to" (OSA13). Similarly, another user liked that on longer, less familiar routes, the GPS provides a way of confirming directions in potentially poorly marked areas: "I like too that on the longer trips it takes me where I want to go. And there are confusing signs here and there on a trip that you take—there may be a sign missing at an intersection" (OSA9). One user implicitly viewed the GPS as reliable when driving through new or unfamiliar places: "I liked it for when we went somewhere that we had no idea where we were" (BSA20). Likewise, another user commented that "the most useful thing is when I'm using it in a place that I'm completely unfamiliar with" (OFT2).

In their usability study of GPS devices, Elliott Noel et al. also note that "[p]articipants were less likely to question the routes chosen when they were in completely unfamiliar places" (380). Indeed, many GPS users interviewed in this chapter viewed the GPS as more reliable and implicitly attributed more agency to the device when driving in unfamiliar settings. Users were quick to note, however, that following the GPS is not always a foolproof method for finding one's way. As one user put it: "[W]e thought it was cool that we could find out where grocery stores or gas stations are that we may not have realized, and then I realized that it didn't always know where things were" (OSA8). Another user described feeling frustrated that the GPS did not always find places relative to their exact location on the road:

> We're going down the road on the interstate, someplace we
> don't know, and we want to have lunch, and you can search
> for food. We have access to the AAA database, which is, I
> don't know, 6,000 entries or something unbelievable, I forget.
> It'll tell us a restaurant that we've *passed*. It would be nice if
> it would ignore those, since it knows which way you're going,
> and it'd be nice if it ignored the way you passed, and tell me
> what's in front of me. Because then alright I have to figure out
> where it is, then I have to take an exit, and it's a pain in the
> neck. It's minor, but it's a pain. It's a pain. (OSA4)

Another user commented on the lack of reliability of the GPS database, even despite recent updates: "[O]n the trip, the type of thing that we used it a lot for was trying to find things like fast food restaurants, you know, grocery stores, and I'd say it was kind of a 50/50 mix; half the time it helped us out a lot in finding where we needed to go, and half the time, its database was just so out of date, even with recent updates" (OSA6). As Mark Monmonier has described, the "GPS calculates location by comparing time signals from several satellites, each with a direct line of sight to the receiver. Each satellite broadcasts a signal travelling at the speed of light but requiring a measurable time to reach the ground" (*Spying* 13). These signals can then be processed by a GPS receiver, "enabling the receiver to compute position, velocity and time" (Dana). While GPS databases allow users to update their devices at regular intervals, this information may not always be fully accurate or the information may be based on outdated paper maps, placing drivers at risk for receiving inaccurate information. For example, in a recent interview aired on National Public Radio's *Weekend America,* Mark Monmonier accounted for a common error that can occur with GPS systems. When asked how GPS errors happen, he said:

> GPS can still be wrong because basically what we are often times doing is taking the coordinates we get from a GPS system and we are putting them on [an] electronic map. And that electronic map might be based on paper maps. Those paper maps might be wrong and there is a possibility that we might have features that were put on the wrong place in paper maps that were otherwise right. So, consequently errors happen. (Cooper and Jefferson)

While many users appreciate what they perceive to be the reliability of the GPS, many users are likewise aware of its various potential limitations and thus do not view it as wholly dependable. Instead, they engage with skepticism in the directions it provides or view it more so as a way to check their directions against other information. At this point, the GPS becomes viewed as less reliable, and users begin looking for ways to supplement the device with different forms of knowledge. Subsequently, the GPS begins to function rhetorically as a supplemental artifact.

The extent and type of supplemental activity performed by the GPS is dependent upon the varied contexts and comfort levels of the particular user. For example, noting some common pitfalls of the GPS, one user recalled their initial apprehension when first using the device: "There was a certain degree of trepidation the first few times using the device. I made sure to print off hard-copy directions from the internet for the first two or three trips, just in case I lost the signal, or the maps on the GPS unit weren't properly updated" (OFM10). A newer user described a similar trepidation and, upon feeling lost, "ended up having to look at a paper map to figure out where I was and where I needed to go" (BA16). Conversely, a more seasoned user commented that the GPS "supports our Google Maps in case we get lost" (OSA8). One user commented on the limitations of both the GPS and internet mapping programs, stating: "I found several 'roads' which don't exist! Of course, I've also found these on MapQuest and Google Maps" (OA12). This user instead discovered and printed out maps from a different source and now keeps them in the car for personal navigation, finding "those maps to be much more accurate than the GPS maps"

(OA12). Leshed et al. also note that "while a few informants never use other navigation aids, others are reluctant only to rely upon the GPS" and also carry supplemental maps with them (5).

Other users prefer to supplement the GPS with paper maps because they feel that paper-based maps more easily afford a broader view of the surrounding areas. These users also implicitly invoke the durability of the paper map; they refer to the ability to carry a map, to pull out a map when needed, or to literally follow a map with their finger. As one user put it: "I'm us[ed] to having a map which shows me a guided path to my destination and allows me a view of my surrounding areas, and that was sort of comforting. With the GPS I realized I would not have that, so that was a little concern and remains a little bit of a concern with the GPS" (OSA9). This user also linked some of this perceived limitation to the smaller screen size of the GPS, as opposed to the less limiting paper map:

> [The] GPS will show you a six inch screen where you are. [. . .] I like to know what's farther around than that, so I will sometimes carry a map so that if I'm going down a route the GPS says, and I want to know what's ten miles left or right, I can pull out the map and see. I don't know if I really have to have that, but that's sort of a comfort, if something would go wrong with the GPS. (OSA9)

Another user also described both a preference for juxtaposing the GPS with the paper map, and an appreciation of the physical tangibility and psychological comfort of the paper map:

> I really like combining [the GPS] with a map. To me, it doesn't replace a traditional map, because you can't see the whole picture—you have a more concentrated area that you're looking at on the GPS, and with a map—I'm very much of a map kind of person—I like to follow exactly where I am. If [name's] driving, I like to literally have my finger on the road and see, oh, yeah, we just passed that town. (OFT2)

Interestingly, this user also describes the ability to follow the map closely, especially while being a passenger in the car. Having both a driver and passenger in the car changes users' interactions with the GPS. This user, for example, notes that "with two people in the car, one person can concentrate on driving, and the other one can be [. .

.] searching up ahead. For instance, [name] likes to drive [. . .] and so when he's driving, he'll say, 'oh, it looks like there's an accident up ahead, check if there's an alternate route,' and so I can be playing with that, while he's figuring out if he should get off there" (OFT2). Leshed et al. too note that "the passenger riding next to the driver sometimes received the role of interacting with the GPS, in terms of switching settings [. . .] or pulling up information" (6). In this way, the GPS engages both the driver and passenger in a type of cooperative supplemental work that not only has consequence on the body but, as I will soon describe, also has broader implications for purposeful decision-making, social relationships, and more nuanced understandings of the environment.

REPRODUCIBILITY, PRESERVATION, AND THE PRACTICES OF MEMORY

The supplemental work performed with and against the GPS functions heterotopically by inviting users to question or resist the cartographic texts it produces, and by extension, as one user described, to question the feasibility of a route's actual reproducibility in the world. For, as Blair writes: "Sometimes what appears to be the rhetorical text is not *the* rhetorical text, but an altogether different one; and what counts as *the* text is open to question" (38–39). In an apt example, one user described cross-checking a potentially confusing GPS itinerary against other directions, not because the directions were wrong per se, but because of different perceptions of what counts as the best route in a given case:

> [W]e'll put in the address, and then from time to time, if I know there's a problematic route that's involved, then I'll actually go and check the itinerary, as opposed to just following it blindly, because when we lived in northern New Jersey for awhile, there was a particular route that it would take us as being the shortest distance between two points, but because it was a fairly low speed limit and quite a lot of lights then it was actually a slower route than taking one of the major highways. (OA1)

In this case, the GPS suggested a route that, while technically encompassing the shortest distance, was, in the user's view, the least feasible to reproduce given local knowledge of the lower speed limit and preponderance of traffic lights. So in this case, the user supplemented local knowledge for the GPS's suggested route. This sort of supplemental activity makes the texts produced by the GPS always subject to the nuances of individual contexts and the changes in reproduction that they invite.

From a Blairian perspective, the GPS text bears various degrees of reproducibility and possibilities for preservation. For example, GPS users may program, save, and thus preserve routes in their list of favorite places and the like. Parks describes the ways in which GPS users "can identify what are referred to as 'waypoints'—significant locations that are meaningful in some way—and use the device to create [their] own landmarks" (212–213). These features are then "figured as a way of remembering one's route" and "may also shape new relations between image and memory" (212–213). Blair also notes that "[t]he link between reproduction of a text and memory is substantial" (38). As Harris and Hazen also describe, it is "useful to think of the relationships that unfold as maps are created, the meanings that are cited in selection of particular technologies [. . .] or as maps are engaged by users" (53). Thus, the ability of the GPS to preserve and reproduce stored routes and saved information, and display this information through multimodal, cartographic texts can indeed foster new schemas for geographic meaning and relationships between the GPS and the user, as well as shape the "new relations between image and memory" that Parks describes. To this end, for example, several GPS users remarked that they appreciated the ability to program and store information, and especially liked the feature that immediately routes the driver back home. One user, for example, described being able to save addresses in the GPS:

> I love how you can save addresses. So if I'm going, you know, to a repeat place, or to some friends, and I can't remember exactly how to get there: [Just] go back in, tap on the address, and it's great. So you don't have to keep reentering, you can actually save, just like on a cell phone, you know, like saving numbers. So that saves time too (OA3).

Another user described liking the "home" feature and the ability to find recent locations: "I thought the fact that you could press the 'Go Home' and type in your home was convenient, or recent locations, that was convenient" (BA22). One user even described the "home" button in terms of its capacity for memory: "It's got a memory button, you can push the 'home,' and it takes you home from wherever you are, you know it routes you there. So once you tell it where home is once, you don't have to do it again" (OSA4). As mentioned earlier, commemorative sites and mapped representations may function as heterochronies that are "linked to slices in time" (Foucault, "Of Other Spaces" 26). They may work rhetorically as memory texts in various ways, for they are not only temporally-sensitive representations of specific cultural moments but they also shift and change given an artifact's timeless interaction with its present context. Understood in this light, the GPS functions here as a different sort of memory text—one that, while not engaged in commemorative work or public memorializing, per se, serves to mediate bodily experiences of remembering nonetheless. The GPS user, for example, can engage in the timely and temporally-sensitive act of navigating home. However, by programming the GPS to *remember* home, the user doesn't "have to do it again." The GPS then functions as a heterochrony that links the timely and present experience of navigating home with timelessness of the accumulating memory of the GPS database.

Users also describe the act of programming and storing information, and being able to easily retrieve information without having to reenter it. As Brooke describes, the "externalization of memory has become an accepted and even integral part of society" (786). He says that the "distribution of memory into the environment" not only means that "'the mind can now keep some knowledge intimately in memory and relegate other knowledge to written texts,'" but it also means that "[i]n many contexts, access to a text is accorded the same importance as knowledge itself" (786). This idea remains important for rhetoric scholars as it is often associated with debates in classical rhetoric about the merit of artificial technologies of remembering. Specifically, the idea is typically associated with Plato's denouncement in the *Phaedrus* of technologies of writing as threatening humans' capacity for being able to recall and recite information from memory, and, as Brooke notes, for its broader implications on "the relation of memory to thought" more generally (784). However, as Brooke writes:

> While Plato could distinguish between natural and artificial memories, the advent of mass literacy seems to put such a debate to rest—at least to the extent that memory loses its most

> important claim to canonical relevance for rhetoric. [. . .] Plato
> may be correct in suggesting that literate peoples will be more
> forgetful, but as a culture, their powers of recall are limited
> only by the material resources they devote to memory. (786)

The GPS indeed allows for the devotion of material resources to memory and functions as a sort of multimodal, digital memory text. It is
likewise implicitly understood as such by its users, many of whom do
accord its address storage and retrieval features the "same importance
as knowledge itself." Brooke also notes that "[a]s our memories and
technologies have become even more artificial, they have done so only
insofar as they circle back and approach the appearance of the natural"
(787–788). We see this case in point, for example, when the one GPS
user describes losing the visual display of the simulated car on the road
as "driving on white" (BA21). That is, this user describes the experience of being off the GPS and driving on white, so to speak, in terms
that sound and maybe even appear just as natural as the surrounding
physical terrain. Subsequently, this user describes having to physically
stop and ask for directions before navigating home. In this account,
we see a toggling between accounts of the technology that "approach
the appearance of the natural" and a return to the bodily experience
that ultimately underpins all GPS use. Brooke feels that we tend to
"celebrate the paradoxical immediacy that our technologies provide
for us, an immediacy that has supplanted the natural" (788). An important caution here and a critical element of understanding the notion of the posthuman body, as expressed by Brooke and Hayles, then,
is to remain mindful of the embodied experience that gives way to the
immediacy provided by the technology (788). In this way, users' discussions of their experiences with the GPS may be understood as implicitly performing understandings of the posthuman mediated body,
through their moving back and forth between descriptions of the immediacy afforded by the technology and the bodily experience that
both underpins and is impacted by its use. It then becomes possible to
see how the GPS not only has consequence on individual, bodily experience but also has broader implications for how we understand our
environments and engage with, describe, or recall the world around us.

THE GPS AND ITS BROADER CONSEQUENCES

Carole Blair writes that "[t]he entry of a text within a particular context is a move on that context that changes it in some way. Perhaps the best way to think about this notion," she says, "is to ask what is different as a result of the text's *existence,* as opposed to what might be the case if the text had not appeared at all" (34). The entry of the GPS within the contexts of everyday navigation signals a shift in the ways that humans have begun to understand and navigate their physical environments. As we have already begun to see, the GPS and its user may co-construct agency in such a way that facilitates purposeful, navigational decision-making. However, the use of the GPS within the contexts of everyday navigation arguably has wider consequences beyond its immediate use to achieve a particular navigational goal while driving, and has a broader potential to impact personal relationships and how we interact with and understand the worlds in which we live.

THE GPS AND ITS INFLUENCE ON SOCIAL
AND INTERPERSONAL RELATIONSHIPS

As Barry Brown et al. describe, the car interior is in many ways a social space: "Sharing a car journey to work (or elsewhere) has the unusual responsibility toward one another which we usually associate with those forms of co-habitation such as sharing a flat or an office" (21). Indeed, just as one GPS user described the ways in which the GPS could impact the sort of cooperative decision-making that happens with a passenger, several participants interviewed in this chapter similarly described with humor but also with great seriousness the ways in which the GPS has somehow affected their social lives or their relationship with a spouse or partner. One user, for example, described the different levels of confidence she and her spouse each bestowed in the GPS and the negotiations that ensued: "I was trying to use it, thinking, you know, it's going to tell us the best way. And so I didn't have to worry about while we were driving, getting out a map and looking, but my husband still wanted me to do that, so he could see where we were" (BSA20). Here, this user again anthropomorphizes the GPS, describing it as able to "tell" them the best route, thereby reinforcing its agent status and seemingly taking on the status of an additional passenger. While this user did not feel that a paper map was necessary, her husband, a fellow passenger, was still interested in supplementing the GPS with other navigational tools.

On the other hand, another user implicitly described the tension that was alleviated by the GPS: "It's definitely been a godsend for my wife. [. . .] [S]he uses it, and it's just been wonderful for her, and for me, because then she doesn't have to call me up for directions. [. . .] so that's been a wonderful feature" (OSA6). One user emphatically described the "intense relief" that accompanied what they described as being able to give over control to the GPS:

> My better half is kind of neurotic about directions, and navigating for him is an experience that is so stressful that I could just not handle it on a regular basis. [. . .] [H]e needed to have directions five steps in advance, and so I would sit there with the map, trying to plot out, like a naval submariner, to figure out where exactly it was that we were going so I could give him [directions]: so in another three miles or so, we're going to need to take whatever exit toward whatever town. [. . .] It was very stressful. [. . .] So it was intense relief, saying that I could give up navigational control. (OA1)

This user interestingly finds relief in being able to afford the GPS with fuller agent status, seeing it as a means of alleviating travel anxiety in a personal relationship. Likewise, another user also described the GPS as helping to lessen the travel anxiety that often preceded a trip and the positive impact of that lessened anxiety on their interpersonal relationship:

> Amusing but true: The GPS has done a lot to help my relationship. [. . .] [D]riving used to be a very stressful activity for me, which left my poor girlfriend with a choice of either being stuck driving everywhere or dealing with the stress of trying to give me directions. With the GPS, she can finally just relax on a trip and not worry about a thing, with the assurance that if I'm stressed or seem angry, it's absolutely nothing to do with her and nothing for her to worry about. That's been a huge boon, since we do tend to go on long trips together, and vacations no longer start with the burden of a huge build-up of tension on the road. (OFM10)

Perhaps most interesting about these accounts is that the GPS functions not only to mediate bodily experience, but by extension, to mediate interpersonal relationships. This mediating function may then have the result of enabling or constraining geographic knowledge.

The GPS, Corporeality, and Geographic Knowledge

For example, one user not only described the GPS as alleviating travel anxiety with a spouse but also described a lessened fear of getting lost and the positive impact of those issues on their relationship:

> I don't have to feel worried about getting lost. Even if it takes me two hours longer to get somewhere, I'm still going to *get* there, or I'm going to get close enough to where I'm going that I can find my way home. And I don't have a good sense of direction, so especially when my husband prints out the Google map and we follow it, but then he forgets to print the return trip and we sit there in the dark, arguing at night: 'No, you need to make a left, no you need to make a right.' [I know that] even though we might not go home the exact same way we came, we can at least get home. And I really like that." (OSA8)

Likewise, another user remarked: "I found it changed my dependency. Normally, I'd just be like, 'Oh, I can find my way,' and I was like, 'Oooh, I like using the GPS, this is easier.' I'm definitely not going to get lost now" (BSA17). In fact, several users commented on the impact of the GPS either on their fear of getting lost or on their changed decision-making while driving. These changes in mindset may impact drivers' navigational decision-making by making them more willing to explore new or unfamiliar places, and thus have potential corporeal impact.

Certainly, keeping in mind the issues mentioned by Monmonier, for example, not all GPS users find the device to be a navigational panacea; as this one user notes: "More than once I've found myself unable to follow directions because the old exit for a given road was a quarter mile or more past the new one, and I wasn't aware what major road I was going to look for [until it was] too late" (OFM10). Similarly, another user commented: "Being stuck in traffic [. . .] on [the highway] due to the perpetual construction and simply having the desire to find parallel roads so that I can exit, actually be moving, and then get back on [the highway] past the construction, I discovered that some of the roads in the GPS aren't really there" (OA12).

Many users also commented, however, that the GPS made them feel more adventurous while driving or made them more willing to make navigational choices that they might not otherwise make without the GPS: "I have the tendency to take familiar routes, and if the GPS tells me that there's another one that is either quicker or goes through an area that's less full of traffic, [then I will take that one.] [. . .] [S]everal of the routes I take to D.C., I'll take because the GPS says that it'll be shorter, and I have changed in that way" (OA1). Another user noted feeling more secure on the road and subsequently noted the positive impact of this confidence on a personal relationship: "It gives me some more confidence on the road, just to be able to go to various places. And my wife likes that because then I'm more adventurous, in being able to go out and do things, and she likes that, so [. . .] yeah, it's been very helpful" (OSA13). Similarly, another user said:

> I think it's made me more adventuresome [. . .] I feel much more confident, as opposed to, as I said, getting directions off the internet, when, that's happened to me before and they're not updated. [. . .] Again, you get off the road and it'll say 're-calculating route,' and it'll tell you exactly how to get back on. So again it's made me more adventuresome. If I want to stop off and look at, like a historic site, I'm not concerned about, 'oh my gosh, how am I going to get back on?' (OA3)

One user also described a newfound ease in making travel plans that accompanied use of the GPS, and the subsequent alleviation of travel anxiety for both the user and passenger: "I quickly became impressed with how much simpler it was making my travel plans, as well as how much less stress it implied for myself and/or my navigator, since missed exits, wrong turns, and road closings were no longer potentially catastrophic failures" (OFM10). Given these examples, then, it is possible to see how use of the GPS would enable changes in social and interpersonal relationships. The GPS may potentially play a role in discouraging the user from exploring the surrounding landscape, or it may help to instill a sort of cartographic confidence that positively impacts social or interpersonal relationships and allows for more willingness to explore the surrounding environment. In either case, the GPS functions as an artifact of visual-material rhetoric that engages both the mind and body and plays an important role in facilitating the bodily experiences of everyday, personal navigation related to geographic decision-making.

CONCLUSION

To use the GPS, then, is to engage with a discursive technology that has real and corporeal impacts for the navigational choices that we make while driving. As a visual-material rhetorical approach helps demonstrate, the GPS and its user engage in an ongoing interactive exchange through which agency is co-constructed by means of multiple rhetorical practices that include: supplementing the GPS and using the GPS supplementally; correcting the GPS and being corrected by it; programming and reproducing routes and challenging those reproductions; engaging in digital memory practices by storing and retrieving addresses and other geographical information; negotiating and challenging its routes through subtle subversions; and engaging with its symbolic representations while remaining attentive to actual physical surroundings. All of these practices have consequences relative to immediate goal fulfillment in the rhetorical situation; that is, these practices are implicated in the driver's presumed interest in learning about or navigating to a destination. However, as is also evidenced by these practices, GPS use has broader consequences on the mediated, posthuman body, beyond those of immediate goal fulfillment. As we have seen, the GPS has the capacity to impact geographical understanding, both through constraining or discouraging geographical knowledge-making and through enabling and encouraging travel and geographical exploration. The potential of the GPS to impact geographical knowledge subsequently affects our everyday, social and interpersonal relationships, and how we choose to explore and understand the worlds in which we live.

To understand the GPS and its user as co-constructing an interactive agency that has the capacity to inform purposeful decision-making is to understand contextualized, mediated bodies from the vantage point of that sort of "can-do-ness" that more explicitly proposes, incorporates, or resists the practices imposed upon them through cultures and social groups. Understood from this perspective, a visual-material rhetorical approach may be seen as illuminating a more informed understanding of the interaction between bodies, spaces, and material discursive practice. In this sense, visual-material rhetoric has the capacity to function in the service of advocacy—not only to help advocate for tangential bodies (as we saw in chapter three), but also, as we have seen here, to illuminate how visual-material artifacts can function rhetorically to enhance informed and technologically mediated decision-making and expand our capacity to understand and explore our worlds.

This more nuanced understanding of visual-material rhetorics as able to function in the service of advocacy should not remain constrained, however, only to its implications for human bodies or technologically mediated posthuman bodies, for that matter. As this chapter has shown, a visual-material rhetorical approach can illuminate the ways in which we approach and negotiate our curiosity about the world around us—a curiosity that need not be constrained only to the consequences of rhetorical action on human animals. As Harris and Hazen note, "[t]he last decade of geographic scholarship has also witnessed a proliferation of discussions related to 'animal geographies,' and linked efforts to rethink human relationships to non-human or 'more-than-human' natures" (51). As Cary Wolfe also notes, issues related to understandings of the nonhuman animal "are only part of the larger issue of nonhuman modes of being" and are thus inextricably linked to how we understand the posthuman (*Animal Rites* 193). Given these views, then, if the nonhuman animal may be understood as part of the broader issue of "nonhuman modes of being," then embodied knowledge need not be constrained solely to human ways of knowing.

Subsequently, to understand embodied knowledge as also concerned with the nonhuman animal means that visual-material rhetorics must likewise be able to account and advocate for nonhuman animal bodies implicated in various discursive settings. With this need in mind, the next chapter works to show how a visual-material rhetorical approach can help advocate for the bodies of nonhuman animals who may be unable to advocate for themselves within the setting of a given rhetorical context. Like chapters three and four, it too demonstrates how understanding the map as heterotopic space can illuminate spatial relationships as rhetorical, and shows how the map works rhetorically given context, purpose, and audience. Chapter five, however, also explores how the map, as an artifact of visual-material rhetoric, can function as a mediating device in policy-making, or as a persuasive tool in debates between institutions. It demonstrates the impact of visual-material rhetorics not just on human bodies, but also on nonhuman animal bodies; it does so by showing how the map can function rhetorically to illuminate competing knowledge claims about contested space in order to influence a contemporary debate about environmental policy and marine mammal protection and advocate for nonhuman bodies who cannot be physically present in the setting of the debate.

5 Advocating for Nonhuman Bodies

On August 7, 2002, the Natural Resources Defense Council (NRDC) filed a lawsuit in a San Francisco federal court against the Navy and National Marine Fisheries Service (NMFS), requesting a preliminary injunction to put a stop to the deployment of low-frequency active sonar (LFA) in areas of the North Pacific Ocean (Calvert and Buck 3). LFA is used to detect submarines in matters of national security and has also shown to be extremely harmful to marine mammals.[1] In October 2002, the court granted the preliminary injunction, and in August 2003, a federal court issued a permanent injunction, restricting deployment of the LFA sonar and ordering the Navy to negotiate with the plaintiffs (the NRDC) the terms of its deployment. In October 2003, the Navy reached a settlement with the NRDC, agreeing to limit LFA deployment, particularly in areas of the North Pacific Ocean.

In August 2002, when the initial lawsuit was filed, a map produced by the NRDC helped persuade the court to grant the October 2002 hearing for preliminary injunction. Of specific interest in this chapter is the October 2002 preliminary injunction hearing, during which the NRDC's map (Figure 46) was invoked in debates over whether the NMFS would deploy LFA sonar in areas small enough to meet "the requirement of a 'specific geographical region,'" as set forth in the Marine Mammal Protection Act (MMPA) (NRDC v. Evans, 2002a 7). Briefly put, the MMPA had strict guidelines that set parameters on any human activity that may result in what the MMPA calls the "taking" (harming, harassing, or killing) of marine mammals; because the deployment of LFA sonar could result in the taking of marine mammals, the NMFS had to show that any incidental takings would occur within the parameters specified by the MMPA's requirements for a specified geographical region.

In this hearing, the NRDC's map was discussed in relation to claims made by the NMFS about the specified geographical region affected by the sonar. In their Final Rule on the Navy's use of LFA sonar (US Dept. of Commerce: NOAA 46768), the NMFS had invoked the Longhurst biome boundaries, which are based on a cartographic representation that divides the world's oceans into specific biomes and provinces (see Longhurst). These areas may then serve to delineate the ocean boundaries in which certain activities may take place.[2] The NMFS had invoked these boundaries to describe where the sonar would be deployed. The NMFS had also made claims about the specified geographical region affected by the sonar, or the area in which takings would occur, in a map they had included in a public PowerPoint presentation about the use of LFA.[3] The map created by the NMFS and included in their PowerPoint presentation may be read as invoking the Longhurst biomes, and was also based on a Mercator projection, common for use in sea navigation, but which made the area of affected ocean appear smaller than it actually was. The NRDC's map of the same area of ocean, an appropriation of and produced in response to the map in the NMFS's presentation, used a different map projection to more accurately represent the much larger area involved (Negin 46). The NRDC's map "was presented at the 8/7/02 hearing, and helped persuade the court to grant the preliminary hearing," which then took place in October 2002 (Luijten, "Referencing"). Together, these maps, although developed and invoked in different contexts, served as the basis for a debate about where the sonar could be deployed, and which representation of the area was more accurate. While the two maps were but one part of the rhetorical situation, their juxtaposition works well to convey the competing claims to knowledge made by the institutions involved, thereby illustrating the importance of the visual-material text.

There were also other variables taken into consideration during the October 2002 motion for preliminary injunction, in addition to the NMFS's alleged lack of compliance with the MMPA, such as whether the NMFS and the Navy violated the National Environmental Policy Act and the Endangered Species Act (NRDC v. Evans, 2002 3). That is, visual-material artifacts rarely work in isolation from other texts and contexts, and the maps in this case functioned as one part of a larger debate over whether the NMFS and the Navy violated several environmental laws through the use of LFA sonar, and in doing so, put at risk the marine mammals residing in those regions of ocean. The role of these two maps in the discussion of whether the NMFS violated the MMPA by misrepresenting the specified geographical region in which the sonar was to be deployed is central to an exploration of how visual-material artifacts illuminate aspects of the rhetorical situation that might otherwise be less apparent. More specifically, this chapter considers how a visual-material rhetorical approach can help advocate for the bodies of nonhuman animals, such as marine mammal species, who may otherwise be unable to advocate for themselves within the setting of a particular debate. As chapter four described, the idea that visual-material rhetorical artifacts can function in the service of advocacy need not remain constrained only to its implications for human bodies. Rather, we may view these ideas as equally relevant to our understandings of the consequences of rhetorical action on nonhuman animals.

ON SELECTION OF THE CASE AND DATA COLLECTION

I first learned of this case during the summer of 2003, after reading an article in a recent issue of *Onearth,* a periodical published by the NRDC. The article dealt with LFA deployment, the court case between the NRDC and the NMFS, and the function of the two maps in the larger debate. At that point, I contacted the NRDC's Geographic Information Systems (GIS) manager at the time. The GIS manager shared with me the map produced by the NRDC in reference to this case[4] and directed me to the NMFS's PowerPoint presentation (available to the public online, see note 3) that contained the map used by the NMFS, to which the NRDC had then responded. While I have been able to confirm that the NRDC's map was presented at the August 2002 request for preliminary injunction, and while the NRDC based their map on the one included in the NMFS's presentation, it should be noted that the map included in the NMFS's PowerPoint presentation is not introduced or mentioned in the Opinion of the Court granting the preliminary injunction or the Summary Judgment.[5] Rather, the Opinion of the Court and the Summary Judgment invoke the NMFS's use of the Longhurst biomes to make claims about the specified geographical regions affected by LFA sonar. That is, in their Final Rule, the NMFS had invoked the Longhurst biome boundaries (US Dept. of Commerce: NOAA 46768–46769), but not an actual map of the Longhurst biomes or the map in their PowerPoint presentation. The NRDC had objected to the NMFS's use of the Longhurst biomes in the Final Rule, as well as the map in the public PowerPoint presentation, and subsequently created their own map in response, which they then presented during the request for preliminary injunction. While I cannot speak to the precise setting in which the NRDC's map was originally invoked and discussed, or the precise statements that were made during that time, I have consulted the Complaint for Declaratory and Injunctive Relief (NRDC v. Evans, 2002); the Opinion and Order Granting Plaintiffs' Motion for a Preliminary Injunction (NRDC v. Evans, 2002a); and the Opinion and Order on Cross Motions for Summary Judgment (NRDC v. Evans, 2003) to better understand the details surrounding the case and the inclusion of the NRDC's map in the preliminary injunction hearing. Most important to the analysis in this chapter is the Opinion of the Court granting the preliminary injunction (NRDC v. Evans, 2002a). It is here that the NRDC's map is invoked and referred to as "Exhibit A" (NRDC v. Evans, 2002a 9), and it is here that the MMPA's requirement of specified geographical region is discussed in relation to the NRDC's map, and in relation to allegations that the NMFS was in violation of these requirements. Again, I should note that neither the Opinion of the Court granting the preliminary injunction nor the Summary Judgment issuing the permanent injunction mentions a specific map produced by the NMFS; rather, they refer generally to the NMFS's use or description of the Longhurst biome boundaries in the NMFS's Final Rule (46768–46769) in making claims about specified geographic region affected by the LFA sonar (NRDC v. Evans, 2002a 9; NRDC v. Evans, 2003 10–13). However, the citation at the bottom of the NRDC's "Authorized Deployment" map (Figure 46) notes that the NRDC's map was based on the one used by the NMFS and included in the NMFS's PowerPoint presentation. Thus, the citation in the NRDC map helps to convey that the two maps were in dialogue with one another at one point in time, albeit in different settings, and made competing claims to knowledge.

This chapter is most interested in these competing knowledge claims and how the maps functioned within the context of a specific rhetorical situation at a specific point in time. I should also note that it is not my intention in this chapter to praise or disparage the practices of any one group or the work they have carried out at any point in time; instead, I am interested in juxtaposing and analyzing specific visual-material artifacts and the work of those artifacts in a particular rhetorical situation. My goal then is to better understand the rhetorical work accomplished by such artifacts and their potential for furthering our understanding of visual-material rhetorics as a sustainable project of inquiry. Before exploring these issues and submitting the maps in this case to a visual-material rhetorical analysis, it is first necessary to briefly review and describe some new ideas about how the map itself functions as an artifact of visual-material rhetoric.

THE MAP AS A VISUAL AND MATERIAL ARTIFACT

As described in chapter one, Turnbull understands maps as rhetorical when he notes that they "have power in virtue of introducing modes of manipulation and control that are not possible without them. They become evidence of reality in themselves and can only be challenged through the production of other maps or theories" (54). Here, Turnbull suggests that maps make knowledge claims through their representations of the environment; as these representations circulate among groups, they accumulate a cultural ethos that may constitute claims to knowledge. Subsequently, those knowledge claims and the ethos they have garnered can only be challenged through the production of new representations or the revisioning of already-existing ones.

Complicating Turnbull's idea is Blair's point that there is a tendency for rhetoric to rely too heavily on the effect of a text's production, or on the relative success of the rhetorical text. Too little emphasis, she notes, is placed on the text's greater consequence, and "when it is addressed at all, it is typically advanced as a reason to study the construction (production values, if you will) of a particular text; and it is frequently understood narrowly as 'success' or goal fulfillment" (Blair 21–22). While we would be remiss not to consider the conditions surrounding the production of these maps and their subsequent relative successes, such as which group was representing particular areas of ocean more accurately and why, we must also remain aware of Blair's point that, while "everyone seems to know that rhetoric is not exclusively about production, and more specifically, that it has consequences that exceed goal fulfillment [. . .] hardly anyone seems willing to address it as anything else" (22). This "anything else," she says, is rooted in materiality. Thus, a case such as this must acknowledge the "material force," or the larger consequences of the two maps in question, beyond solely the "goals, intentions, and motivations of [their] producers" (22). Additionally, as chapter three also described, it is not always possible to know or understand the intentions or motivations of the author, especially when they are affiliated with large institutions or government agencies, as is the case with the maps produced by the NMFS and NRDC. While it is necessary and helpful, then, to understand the conditions of the map's production, it becomes even more important to understand the larger consequences of the map, beyond its initial goal fulfillment in the most immediate and visible aspects of the rhetorical situation.

With these ideas in mind, this chapter analyzes the work of two maps in and around the same rhetorical situation. I combine Foucault's concept of heterotopias with an extension of Blair's approach to material rhetoric, in order to conceive of the map, traditionally understood as a predominantly visual artifact, as bearing material qualities as well. I suggest that maps, as heterotopias, may be linked to "slices of time," or heterochronies, at specific cultural moments during which there is a perceived need for social or environmental change (Foucault, "Of Other Spaces" 26). Likewise, the maps related to this case purport to represent a very specific area of the North Pacific Ocean during a specific slice of time. These maps were each produced in response to the Navy's testing of LFA sonar at a particular cultural moment, during which it was becoming more understood that marine mammal populations were vulnerable to LFA sonar. The maps were not developed in response to a general feeling or perception of crisis; rather, they were produced in response to specific events related to the potential harming of marine life through the use of LFA sonar, and consequently, one group's perceived need for environmental change. In this way, the maps involved in this case each "frame and limit the specific geographies and management opportunities possible in terms of how human and more-than-human worlds are inhabited and lived" (Harris and Hazen 63).

The maps also each invoke specific visual, cartographic conventions such as color, projection, scale, and the use of lines and numbers in order to represent a section of the North Pacific Ocean in a particular way, and are thus "reflective of, and productive of, power" (Harris and Hazen 63). These ways of knowing not only tap into existing cartographic design conventions specific to a certain group of specialists, but also give the maps a material component, or varying degrees of what Blair might refer to as "durability," in terms of the extent to which each is perceived as making viable, durable claims to knowledge, and the ways in which their textual composition aids in making these claims. Blair's approach to analyzing objects of material rhetoric then also provides a helpful framework for describing the rhetorical consequences of visual artifacts such as the map in social contexts. I compare the two maps related to this case in order to show how each communicates a subtly different message about the geographic area subjected to the sonar, and consequently, it would follow, the potential harm to marine mammals residing in the regions affected by the sonar. And once again, to describe how each map shapes and communicates a different message also pertains to how the map works in social contexts—an idea that Blair's approach helps illuminate.

A visual-material rhetorical approach helps demonstrate how different spaces, depending on their surrounding contexts and purposes, will have different consequences on the bodies that inhabit them. While Blair speaks primarily of physical spaces, I suggest that we may extend her theory to also account for visual representations of space such as the map. This is because, as discussed in chapters three and four, while maps are traditionally perceived as visual artifacts, they have a material component as well. In extending Blair's framework for material rhetoric, then, we might also argue that: 1) maps are visual texts whose existence may influence our understanding of a place and, as we will see in this chapter, may also shape policy and decision-making; 2) they exhibit various degrees of durability, both in terms of their composition and in terms of the perceived feasibility of their knowledge claims; 3) they bear various degrees of reproducibility and possibilities for preservation; 4) they work with and against other maps to form knowledge claims—a large focus of this chapter; and 5) they act on the bodies residing in the spaces which they represent. Accordingly, this chapter understands the map as a powerful artifact of visual-material rhetoric—one that, in this case, ultimately plays a role in the shaping of environmental policy through its challenging of one group's claims to knowledge.

While chapters three and four respectively described how the sculptures and green spaces at the Lowell Mills National Historical Park work with and against each other to participate in a rhetorical situation, and how the GPS and its user work together to co-construct the conditions necessary for purposeful geographic, navigational deci-sion-making, this chapter more explicitly describes the ways in which artifacts such as the map can function rhetorically to make competing knowledge claims about contested space—claims that may then play a role in the shaping of environmental policy and advocate for nonhu-man animal bodies. It shows how the knowledge claims inherent in each map depend on the features of the map itself and their under-standing of context. Ultimately, it shows how the map is a visual-ma-terial rhetorical artifact that indeed affects the bodies of those residing in the heterotopic mapped space.

THE MAP AS A HETEROTOPIC REPRESENTATION OF SPACE

For Foucault, heterotopic spaces are heterogeneous, sometimes con-tested sites that "always presuppose a system of opening and clos-ing that both isolates them and makes them penetrable" ("Of Other Spaces" 26). These sites, he says, tend not to be "freely accessible like a public place. Either the entry is compulsory [. . .] or else the indi-vidual has to submit to rites and purifications" (26). Other entries may appear "pure and simple," he says, but may "hide curious exclu-sions," and entry here becomes but an illusion (26). Notable here is the idea of exclusions; that is, maps require that we tap into existing cultural norms and symbols in order to extract their meaning. They are selective—they provide clear points of entry into understanding certain aspects of the environment, while excluding other features of the terrain. Moreover, as described in chapter two, if Foucault suggests that places such as the cinema or the theater may be conceived of as heterotopic sites, then heterotopias may be understood as participating in visual culture. If heterotopias participate in visual culture, then I suggest that we may also conceive of visual texts such as films or maps as heterotopic artifacts. Maps work well in this configuration, as they purport to represent real places and are often concerned with con-tested representations of space. Lastly, I suggest that heterotopias may be understood as rhetorical—as visual-material textual sites that have the potential for consequence and can influence our lived experience. In this view, conversely, rhetoric may be understood as concerned with space and place—an idea illuminated through an analysis of the maps used by the NMFS and NRDC.

AN ANALYSIS OF THE MAPS USED
BY THE NMFS AND NRDC

The map used by the NMFS was included in their public Microsoft PowerPoint presentation related to the Navy's request at the time to deploy LFA sonar in areas of the North Pacific Ocean. The MMPA defines "take" as: "to harass, hunt, capture, collect, or kill, or attempt to harass, hunt, capture, collect or kill, any marine mammal" (NRDC v. Evans, 2002a 6). In order to receive a "small take authorization," any activities had to be restricted to a "specified geographical region," result in the taking of a small number of marine mammals, and have only a "negligible impact" on the species (6). Further, "specified geographical region" is defined by the Federal Code of Regulations as "'an area within which a specified activity is conducted and which has similar biogeographic characteristics'" (8). Such a region needs to be "narrowly identified so that the anticipated effects will be substantially similar" (8). That is, for example, it would be "inappropriate to identify the entire Pacific Coast of the North American continent as a specified geographical region, but it may be appropriate to identify particular segments of that coast having similar characteristics, both biological and otherwise, as specified geographical regions" (8).

The map in the NMFS's PowerPoint presentation also included the title and subheading, "MMPA Small Take Authorization Determinations: 1) Will incidental takings occur in a specified geographical area? *Yes*" (Hollingshead 7). Thus, the map made the explicit claim that any takings, or harm to marine mammals, would occur only within a specific region of ocean. The visual aspects of the map would presumably then serve to further support this declaration, visually depicting the areas of ocean to be affected by the sonar. The combination of these textual and visual claims function as what Wood and Fels refer to as a "system of propositions" that consists of a "posting" ("Don't Skip" xvi). Postings, they feel, not only serve to construct the territory in the map but also help account for the map's claim to authority (xvi). A posting, they say, is

> the fundamental cartographic proposition that *this is there.* Each posting encapsulates a powerful existence claim—*this is*—that gains enormous power by being posted. [. . .] Multiple postings participate in the construction of a territory, which facilitates the transmission of authority. [. . .] What

gives the posting its uniquely powerful ability to make existence claims is the social assent that is given to the proposition—this is there—that it embodies. [. . .] The assent given
to the postings spreads to the territory that the postings mutually construct, and this endows the map with an intrinsic
factuality whose social manifestation is the authority the map
carries into public action. (xvi)

In this case, then, the initial authority of the NMFS map not only
comes from its proposition that incidental takings, or a small take,
would occur "there," so to speak, but also in the social authority granted to the genre of the map generally, over time and throughout culture.
Thus the map may be understood as functioning heterochronously, in
that it bears a timeless authority that also taps into and is shaped by
the timeliness of the particular cultural moment that constitutes each
contextualized instance of map-making.

Considering more specifically its visual design components, the
NMFS map may be understood as both iconic and symbolic. It is
iconic because it bears a likeness to the physical territory it represents,
that of the North Pacific Ocean and its surrounding landforms. The
map in the PowerPoint presentation depicts the North Pacific Ocean
as centered, making it the focus of the visual, while the surrounding
coast is present in order to provide perspective (Hollingshead 7). The
map is symbolic because it makes use of conventional symbols such
as lines and numbers, and graphical devices such as the use of color.
In the map, the ocean is divided into "provinces," or "biogeographic
areas," consistent with the Longhurst biomes. Each area is referred to
by its name and a number between 50–70, and is also assigned a color.
For example, Province 56, pictured at the center of the map in light
blue, is referred to as the North Pacific Tropical Gyres West. Province
60, to the right of province 56 and shown in light purple, is referred
to as the North Pacific Tropical Gyres East (Hollingshead 7). Because
the map does not provide the actual area of each province, however,
readers must speculate as to which areas are the largest provinces to be
affected by the sonar.

Readers must also speculate as to the rationale for particular uses
of color in the NMFS map. That is, one might assume, upon a first
reading, that the colors used in the map would indicate which depicted areas of ocean will be most affected by LFA sonar. Or, the viewer
may infer that all provinces depicted are to be affected by the sonar,

and that the color is employed simply to distinguish between province boundaries. There is no legend in the map and no discussion in the surrounding presentation, however, that explains the use of color in this particular map.[6] The claim that only limited areas of ocean will be affected by the LFA sonar would be seem to be consistent with the highlighting in light red, thus drawing greater attention to the smaller, coastal Provinces 53 and 67. Edward Tufte suggests that red is often used to indicate a higher level of importance than other colors: "Color often generates graphical puzzles. [. . .] The mind's eye does not readily give a visual ordering to colors, except possibly for red to reflect higher levels than other colors" (154). However, because there is no legend or no indication of the size of each province, it is not possible to know for certain to which areas these colors refer, and thus it is not possible to make distinct connections between the use of color and the intended deployment of the LFA sonar in this map. Viewers may infer more generally that the areas depicted in the map not only represent the ocean's biogeographic areas but also bear some relation to those areas to be affected by the sonar.

The significance of this map's existence, as it may be inferred by its viewer, lies in its textual claims that only specified areas of ocean will be affected by the LFA sonar, and thus that any takings will occur within that specified area. These claims are made explicit through the map's title and subheading: "MMPA Small Take Authorization Determinations: 1) Will incidental takings occur in a specified geographical area? *Yes*" (Hollingshead 7). However, it is difficult for the map to make this claim with clarity. This is because the NMFS's map is based on the Mercator projection, which makes it less ideal for calculating area. While "no one projection is the best or the most accurate," different types of projections have different purposes for which they are more or less well-suited: "A particular projection is selected by the mapmaker on the basis of functional and perhaps aesthetic criteria, or because of a specification or convention" (Turnbull 6). Mark Monmonier has a more overtly rhetorical view of map projections, especially when describing distortions resulting from the Mercator projection, which he feels to be an unsuitable choice for any map "not related to navigation" (*Mapping it Out* 53). The Mercator projection (Figure 45) is best used in matters of sea navigation, because the "projection vastly enlarges poleward areas so that straight lines can serve as [. . .] lines of constant geographic direction" (Monmonier, *How to Lie with*

Maps 94). In other words, any straight line, or rhumb line, drawn on a Mercator projection represents the actual compass bearing. Because the Mercator projection greatly distorts area, however, its use is generally understood as questionable in matters regarding the estimation of area.

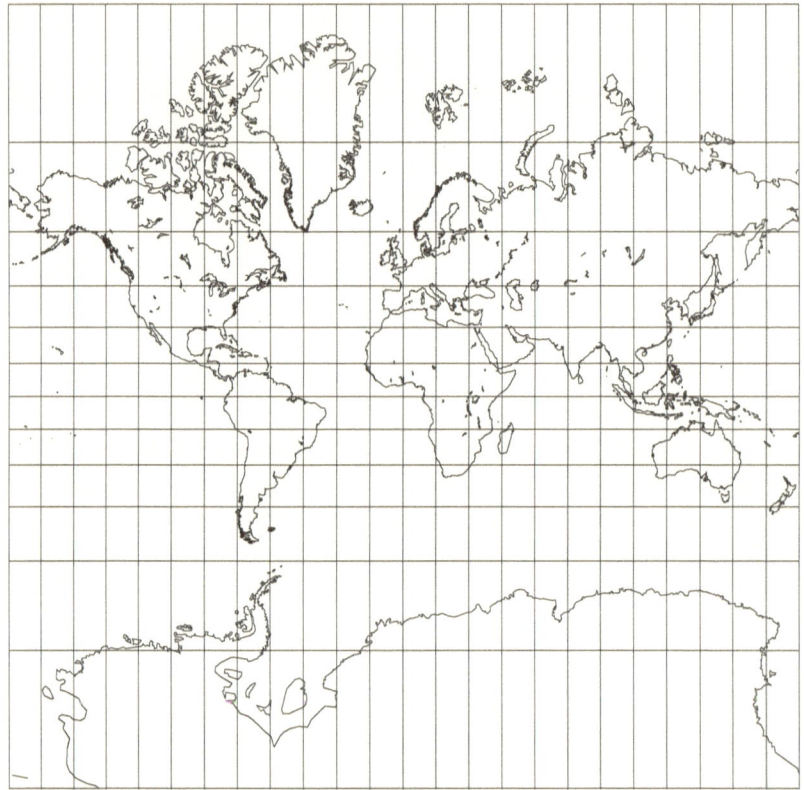

Mercator projection; Cylindrical; Conformal; Gerardus Mercator; 1569

Figure 45: Mercator Projection. Paul B. Anderson, 2002.

The use of color and other visual cartographic conventions such as projection may be understood as contributing to the durability of the map's composition and subsequent knowledge claims. That is, Blair describes a text's composition as contributing to its "durability." A text's materiality, she says, "varies in both degree and kind"; subsequently "the kind of material the text is composed of must be a serious consideration" (37). We might then understand cartographic design convention as one component of a map's materiality, especially given its potential to affect the bodies represented through it. The NMFS map's textual content also begins to convey its broader, material consequences. That is, the map's title, "MMPA Small Take Authorization Determinations: 1) Will incidental takings occur in a specified geographical area? *Yes*," conveys that "incidental takings," or harm to marine mammal life, would be acceptable if contained within a specified geographic region (Hollingshead 7). Understood as such, incidental takings are implicitly viewed as a result of controllable human actions, an idea that arguably reveals a hierarchical relationship between humans and nonhumans and the impact of human action on marine mammal bodies. The content and design of the title, and the use of red font in the subtitle and the underlining of the word "Yes" in particular, may be read as constructing and perpetuating an implicitly binary relationship between human/nonhuman, or, as Haraway describes, one of Latour's "Great Divides," [7] or binaries of "animal/human, nature/ culture, organic/technical, and wild/domestic," all of which she says "flatten into mundane differences" rather than foster more nuanced understandings and embodied ways of knowing (15; 20–21).

The choices related to which features of the terrain are included and excluded from the map in the NMFS's presentation should pertain to the fact that the NMFS wanted to depict which provinces would be affected by deployment of the LFA sonar. It would then be necessary for their map of the North Pacific Ocean to make these provinces visible while excluding other possible features. Such choices might then make clearer the implicated provinces, calling them out through a contextually descriptive use of color. Arguably, then, there exist some interpretive challenges with the NMFS map that manifest when viewing it both in the context of its intended use and, as we will see, in comparison with the NRDC's version of the map. That is, the map in Figure 46, produced by the NRDC in response to the NMFS's map, displays a different understanding of context and purpose and subsequently a different assessment of potential impacts on marine mammal life.

The map in Figure 46 was produced by the NRDC, and as the note below states, is based on the map in the NMFS's PowerPoint presentation:

> NRDC produced this visual representation from the following information:

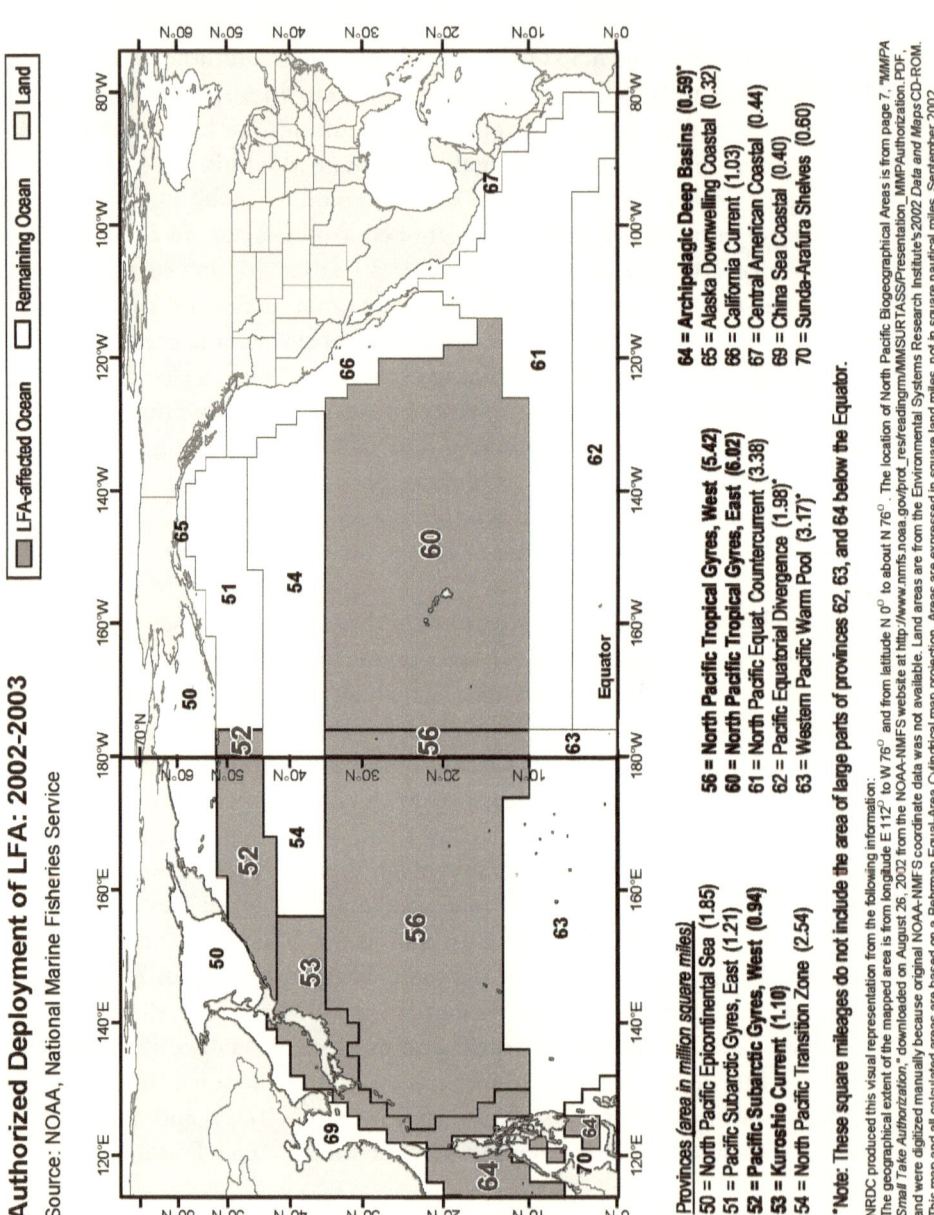

Figure 46: Authorized Deployment of LFA: 2002–2003. (Courtesy NRDC).

The geographical extent of the mapped area is from longitude E 112º to W 76º and from latitude N 0º to about
N 76º. The location of North Pacific Biogeographical
Areas is from page 7, 'MMPA Small Take Authorization,' downloaded on August 26, 2002 from the NOAA-
NMFS website at http://www.nmfs.noaa.gov/prot_res/
readingrm/MMSURTASS/Presentation_MMPAu-
thorization.PDF, and were digitized manually because
original NOAA-NMFS coordinate data was not available. Land areas are from the Environmental Systems
Research Institute's 2002 Data and Maps CD-ROM.
(NRDC 2002)

Given the NRDC's reproduction of the NMFS's map, it is important
to note Blair's point that "[r]eproduction is an intervention in the materiality of the text, and it is important to grapple with the degrees and
kinds of change wrought by it" (38). Similarly, Turnbull has noted
that the map "can only be challenged through the production of other
maps or theories" (54). And as noted in chapter one, Harris and Hazen
also acknowledge "the multiple, reiterative production and reproduction of maps as they are engaged in multiple times and spaces" (51).
Thus, to challenge the knowledge claims put forth by the NMFS, the
NRDC felt it necessary to create their map based on the version included in the NMFS's public PowerPoint presentation.

While the two maps share many visual characteristics, there are
several important differences between them. Most immediately apparent is the NRDC's different use of color. As noted in chapter one, Harris and Hazen also understand color as an important component of
how maps shape meaning, acknowledging the "key insights [that] are
possible by analyzing the ways that lines and colours *become* maps, are
given meaning, and are performed in relation to specific knowledges"
(51). In the NRDC's map, provinces corresponding with areas of LFA-
affected ocean are highlighted in bright red, and a legend then indicates these areas as "LFA-affected Ocean" (*Authorized Deployment*).
While the NRDC's use of red may simply reflect their in-house style
guide for cartographic convention, the color red, as mentioned earlier,
may also be viewed within Western culture as symbolic of warning; it
is therefore possible to infer from this map and its legend the idea that
LFA-affected ocean may be viewed as a threat to the bodies residing
in those areas. According to geographer Denis Wood, the legend func-

tions as a sign system: "[T]he role of the legend is less to elucidate the 'meaning' of this or that map element than to function as a sign in its own right. [. . .T]he legend refers not to the map (or at least not directly to the map), but back, through a judicious selection of map elements, to that to which the map image itself refers" (Wood 101). Thus, given Wood's ideas, we may understand the legend in the *Authorized Deployment* map as a sign that refers back to the idea of LFA-affected ocean in some capacity. The legend in the NRDC map does not visually represent the biogeographic regions in physical, geographical terms, as the map itself does; rather, it represents LFA-affected ocean as a graphical feature associated with the color red.

We might also recall that it is not possible to discern from the NMFS's map the actual size of each biogeographic region in the North Pacific Ocean. The NRDC's map, on the other hand, while also depicting the biogeographical regions, provides the actual area, in million square miles, for each province. For example, the NRDC map shows Province 56 to be 5.42 million square miles, and Province 60 to be 6.02 million square miles. This information, coupled with the information conveyed by the legend, allows the reader to interpret that provinces 56 and 60 are the largest areas to be affected by the LFA sonar. That it is not possible to discern from the NMFS map its area in square miles is potentially problematic, given the purpose for which the map was used. As Monmonier notes, while the Mercator projection is a "marvelous invention for sailors, because it shows lines of constant direction [. . .] as straight lines, [this] equatorially centered [. . .] projection [. . .] so grossly distorts area and distances that the poles lie off the map at infinity" (*Mapping it Out* 48).

The NRDC, on the other hand, based their version of the map on the Behrmann Equal-Area projection (Figure 47), which is "in many ways the opposite of the Mercator projection: (1) distances between parallels decrease towards the poles; (2) the [. . .] projection is not suitable to measure compass bearing; and (3) areas are maintained properly anywhere on the globe" (Luijten, "Greetings").

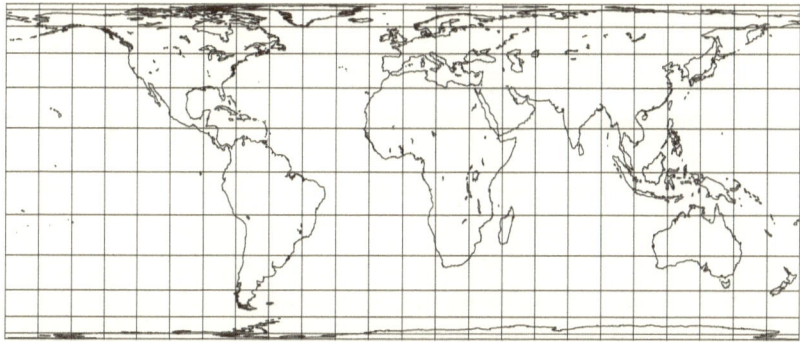

Behrmann projection (Modified Cylindrical Equal-area); Standard Parallels = 30 Deg. N/S; Walter Behrmann; 1910

Figure 47: Behrmann Projection. Paul B. Anderson, 2002.

In looking at the NRDC's map, it is possible to see that the distance between parallels indeed decreases toward the poles; it is visually apparent that the area of the North Pole appears compressed in this map, as a result of the use of the Behrmann projection. The NMFS's map, based on the Mercator projection, appears the opposite; the area toward the North Pole appears taller than, or at least as tall as the area of ocean shown in the rest of the map. While the NRDC could have countered the NMFS's map by making their own Mercator-based map that only revised aspects such as the inclusion of a legend, or the use of color, they understood the choice of projection as an additional component of how maps make meaning, and as such, chose to use a different projection in their appropriation of the NMFS's map. This understanding allowed the NRDC to create a more durable map given its purpose, and likewise make more durable claims to knowledge. Thus, the NRDC's map differs from the NMFS's, in terms of its allowing for more intuitive interpretation, and through its more suitable use of projection, given the context and purpose of the map.

While the NRDC's map also implicitly reveals a hierarchical relationship between humans and nonhumans by mere fact of the work it must accomplish in countering the claims in the NMFS map, the title of the NRDC's map does not explicitly refer to or assume incidental takings. Rather, the NRDC map refers only to authorized deployment of LFA from 2002–2003. Interestingly, its legend refers to "LFA-affected ocean," opposing it to "remaining ocean." When viewing the "remaining ocean" in relation to the "LFA-affected ocean," the reader may then infer that relatively little unaffected ocean will remain given the areas authorized for LFA deployment in the map. Subsequently, readers may then infer a large impact on marine mammals residing in or migrating through areas of LFA-affected ocean. However, as I will soon discuss in more depth, such connections are not explicitly stated in the map.

It is now apparent that within these two maps we have two contradictory versions of a single territory. Foucault says that "[t]he heterotopia is capable of juxtaposing in a single real place several spaces, several sites that are in themselves incompatible" ("Of Other Spaces" 26). These maps then constitute competing heterotopias of this region of the North Pacific Ocean. By comparing and contrasting these two heterotopic spaces, we are able to see how each makes subtly different claims to knowledge, and how together the maps constitute contested versions of a place. In other words, juxtaposing two incompatible versions of a single, real place sheds light on the differing spatial relationships and the consequences of each, and thus helps support an understanding of those differences as rhetorical.

The NRDC's map is able to make durable claims to knowledge through its use of color, its inclusion of a legend, and through its work with and against the NMFS map. Subsequently, the map's persuasive power ultimately earned it recognition in the context of the federal court case in which it was immersed. At the same time, to suggest that the NRDC's map saved the day, so to speak, not only paints an overly simplified portrait of how these maps functioned in the rhetorical situation but also focuses too narrowly on the idea of immediate goal fulfillment. As mentioned at the outset of this chapter, the NRDC's map functioned as but one part of a larger courtroom debate over whether the NMFS violated several environmental laws through its deployment of LFA sonar. It is at this point that it becomes helpful to acknowledge the broader "material force" of the two maps in question, beyond solely the "goals, intentions, and motivations of [their] producers" (Blair 22). And sometimes, the broader material consequence of the text has to do not with its force, but instead with its limitations in accounting for all aspects of the rhetorical situation.

LIMITATIONS OF THE MAP AND THE BROADER CONTEXT

Ostensibly, the main purpose of the NRDC's map was to point out the alleged misrepresentation of area of LFA-affected ocean in the

NMFS's map, and in doing so, point out the potential threat to marine mammals within that area; and with their map, they were able to do so. However, further examination of this case reveals that there exist two main problems in the cartographic representation of a specified biogeographic region. First, the number of million square miles affected by the sonar needs to be represented accurately. Again, the NRDC was able to represent this through their use of the Behrmann Equal Area projection. Second, however, the map needs to be able to account specifically for LFA impact on marine mammal populations. That is, while the NRDC was able to account for the issue of area by choosing a different map projection, their map still could not explicitly account for the second problem, which involves how to represent a specified biogeographic region specifically with regard to marine mammal populations. The Opinion and Order Granting Plaintiffs' Motion for a Preliminary Injunction helps describe this issue. At the time of the preliminary injunction hearing, the biogeographic regions delineated by the NMFS in their Final Rule

> divid[ed] the oceans into 15 biomes, and 54 provinces within those biomes, as designed by Longhurst (1998). [. . . The] NMFS stated that it believed that this approach met the statutory definition [of specified geographic region] because "a biome is the most likely geographic region to contain the majority of a specific marine mammal stock, especially those that are migratory." While admittedly, the Longhurst schematic was designed for plankton, it is the best scientific application available for designating specified geographic regions because no biogeographic concept has been designed for marine mammals. (NRDC v. Evans, 2002a 8)

Here the Opinion and Order granting the preliminary injunction invokes the NMFS's use of the Longhurst biomes in the Final Rule. The NMFS says that although the Longhurst schematic was not designed specifically with marine mammals in mind, they feel that their method is the best available option for designating specified geographic regions relative to marine mammal populations because no such concept exists for marine mammals. During the preliminary injunction hearing, the NRDC had invited a marine ecology specialist to counter the argument that the NMFS's use of the Longhurst schematic was appropriate for designating specified geographic regions relative to marine mammal populations. While the declaration of the marine ecology specialist could not be considered by the Court due to "reasons stated in its Order on Parties' Requests to Strike Extra Record Documents Submitted in Connection with Cross-Motions For Summary Judgment" (NRDC v. Evans, 2003), he had stated that "the Longhurst biomes are not particularly useful for estimating biological impacts on specific populations of marine mammals or other organisms" (qtd. in NRDC v. Evans, 2003 11). More specifically, he stated that

> the provinces identified by NMFS are so large that each one contains many diverse habitats, species assemblages, and levels of productivity. [. . .] "Even if NMFS' purpose in creating very large biogeographical provinces was to ensure that they contain whole stocks of migratory marine mammals, the boundaries are somewhat biologically arbitrary, failing to correspond to population distributions of gray whales, blue whales, and other species." (NRDC v. Evans, 2002a 9–10)

According to the marine ecology specialist, then, these areas were too large to measure impacts on specific marine mammal populations.

During the preliminary injunction hearing, the NRDC objected to the biogeographic regions delineated by the NMFS. According to the Opinion and Order granting the preliminary injunction, the plaintiffs (NRDC)

> object that the biomes and provinces identified by NMFS are still far too large. Plaintiffs have provided a map, *attached as Exhibit A to their motion,* showing the very large size of some of these provinces. According to plaintiffs, Province 60 is larger than the continental United States and encompasses six million square miles of open ocean. The Court notes that Province 66 covers the entire Pacific coast from roughly Cabo San Lucas at the southern tip of Baja California to the Canadian border. Plaintiffs argue that if "it would be inappropriate to identify the entire Pacific coast of the North American Continent as a specified geographical region," [. . .] then surely an area twice the size of the United States violates the MMPA. (NRDC v. Evans, 2002a 9, emphasis added)

The NRDC introduced their map (Figure 46) as "Exhibit A" in the preliminary injunction hearing in order to point out the large size of the provinces affected by LFA sonar. The function of the map's introduction as evidence may also be understood as preserving it in time, remaking it, or creating a new heterochrony of sorts—one that is defined by its submission as "Exhibit A." Blair notes that photographs (or maps, for our purposes) cannot match the experience of actually seeing or being in a place; rather, they "two-dimensionalize and freeze an experience of three dimensions and movement, accommodating a kind of sharing of experience, but only a limited kind" (38). Harris and Hazen acknowledge the "reiterative processes through which map meanings and effects are constantly remade," noting that through these processes, "relations *appear* as stable or natural, even as they are constantly unfixed and remade" (54). The map's ability to freeze an experience of a place or create a sort of timeless representation while simultaneously creating a timely instantiation of a place that is limited to the contextual moment affords the map with a heterochronous property. In this case, the ability of the map to take on the designation of "Exhibit A" allowed it to be understood by the federal court as bearing specific relevance to the rhetorical situation. Blair says that memorials,

> by their presence, do something closely akin to what has been called the agenda-setting function of the televised news; because a topic appears on the news, it is *thereby* deemed newsworthy. [. . .] Similarly, when a memorial (or any other text) appears on the landscape, it is thereby deemed—at least by some, and at least for the moment—attention worthy. (35–36)

I suggest this is precisely the case with the introduction of the NRDC's map as Exhibit A. It appeared on the landscape of the federal court, so to speak, and having been given the ethos of exhibit status, it was thereby deemed attention-worthy, at least by some, at least for the moment. However, it is important to note that while the map indeed plays a significant role in shaping the court's perception of what counts as an accurately portrayed biogeographic region, the map has limitations as well.

The NMFS invoked the Longhurst biomes because no other alternative biogeographical scheme existed in which the specified biogeographic region also defines an area within which specific marine mammal populations reside. The NRDC, while unable to account for this issue with their map, was able to more accurately assess the number of square miles comprised by each specified biogeographical region through their use of the Behrmann Equal Area projection. By more accurately portraying the size of the affected areas, it would follow that those areas will more accurately represent marine mammal populations within each region. Because there was no definition of "specified geographic region" specific to marine mammal populations, however, the NRDC was unable show this correlation directly. Even though the NRDC was better able to represent the area of LFA-affected ocean, the broader implication of their map (that this representation will subsequently allow for accurate measurement of impact on marine mammals) is difficult to make wholly explicit, largely because there was no legal precedent for biogeographic regions specific to marine mammals. Thus, the Opinion and Order granting preliminary injunction states:

> Plaintiffs [the NRDC] have established serious issues with respect to whether NMFS violated the MMPA by choosing such undifferentiated geographical areas, particularly in light of the failure to carve out sufficient areas of special biological importance for feeding, breeding, and the like that lie within these large areas and make them less homogenous. [. . .] Plaintiffs have not presented any evidence, however, disputing [the] NMFS's conclusion that no alternative biogeographical scheme currently exists for marine mammals that can readily be applied here. [. . .] Although the NMFS' choices may be flawed, on this record they do not appear to be so flawed that the Court will likely invalidate them as arbitrary and capricious. At most, plaintiffs have raised a serious question on the merits. (NRDC v. Evans, 2002a 10–11)

Ultimately, the NRDC was successful in presenting a road block in the Navy's ability to deploy LFA sonar. That is, the Court eventually ruled that

> plaintiffs have shown that they are likely to prevail on establishing violations of the MMPA, NEPA [National Environmental Policy Act], ESA [Endangered Species Act], and the APA [Administrative Procedure Act]. [. . .] Plaintiffs have also shown the possibility of irreparable harm to the marine envi-

ronment that supports the existence of these species. (NRDC
v. Evans, 2002a 37)

Here it becomes possible to see more explicitly the consequence of the
visual-material text on marine mammal bodies; that is, the NRDC's
map played a role in convincing the court that the LFA sonar posed a
threat to marine mammal populations. Due largely to the durability
of its knowledge claims, the NRDC was able to show "the possibility of irreparable harm to the marine environment that supports the
existence of these species" (NRDC v. Evans, 2002a 37). Accordingly,
the Court granted the preliminary injunction, which resulted in the
restriction of the sonar's use in particular areas of the North Pacific
Ocean, and ordered both parties "to meet and confer on the precise terms of a preliminary injunction consistent with this opinion"
(NRDC v. Evans, 2002a 39). In the interim, the Court stated that
"defendants should not deploy LFA sonar" (39). Following this case,
in August 2003, a permanent injunction was issued, which restricted
the Navy's use of LFA sonar.

While the Court agreed that the NRDC raised serious questions
with regard to whether the NMFS violated the criteria for specified geographic region as set forth in the MMPA, it did not feel that the NRDC
presented a solution to the entire problem. That is, the NRDC's map
arose in response to their idea that the map in the NMFS's PowerPoint
presentation misrepresented the area of LFA-affected ocean through
its depiction of ocean biomes set within a Mercator projection. The
NRDC also objected to the use of the Longhurst biomes generally, as
set forth in the NMFS's Final Rule, stating that they were too large to
determine biological impacts on specific populations of marine mammals. While the NMFS may not have fully represented the area of
LFA-affected ocean or its potential impact on diverse marine mammal
populations in these contexts, it is also necessary to note that there was
no precedent for a schematic that both accurately portrayed the ocean
area and also accounted for the marine mammal populations therein.
Thus, there were limits to just how much of the situation the NRDC's
map could account for, and it also becomes possible to see the rationale
for the NMFS's reliance on the Longhurst biomes in communicating
the areas in which the sonar would be deployed. It is at this point that
we see not only the merits of the visual-material rhetorical artifact but
also its limits and contextual nuances as well.

CONCLUSION

This analysis has focused specifically on the idea of the map as a persuasive visual-material artifact that can help illuminate the tensions and nuances of a court case whose outcome influenced the shaping of environmental policy and the lives of the marine mammal bodies residing in the space represented by the map. Here, the theories of Blair and Foucault serve as stepping stones that provide access to the more important question of how a visual-material rhetorical approach can afford a window into the larger consequences that these artifacts have in the world. We have now seen how maps can be understood as visual-material and heterotopic rhetorical artifacts. By first understanding maps as heterotopic spaces, and next by juxtaposing the NMFS and NRDC maps and their function in various contexts, we may see how each makes subtly different claims to knowledge, and how together the maps constitute contested versions of a place. Ultimately, Foucault's theory of heterotopias is helpful because it allows for the juxtaposition of two incompatible versions of a single, real place, in order to illuminate the differing spatial relationships and understand those differences as rhetorical. It is also helpful because by juxtaposing the two heterotopic representations, it is possible to see their rhetorical consequences for the bodies represented by them.

As an artifact of visual-material rhetorics, the NRDC's map functions as a text whose existence helped influence the federal court's understanding of the rhetorical situation, such that they restricted the use of LFA sonar in areas of the North Pacific Ocean. The NMFS and the NRDC's maps exhibit various degrees of durability, both in terms of their composition—the digital format of one enabled the digital reproduction and appropriation of the other—and in terms of the relative durability of the knowledge claims of each. As Blair's theory of material rhetoric helps describe, the NRDC's map, submitted as "Exhibit A," deemed it "attention-worthy" in the eyes of the court and granted it ethos as presumed fact (35–36). Likewise, as Latour has noted: "A text or statement can thus be read as 'containing' or 'being about a fact' when readers are sufficiently convinced that there is no debate about it and the processes of literary inscription are forgotten" (Latour 76). Perhaps, then, the map's designation as "Exhibit A," and the cultural authority implicitly granted through such a designation, functioned to override the processes of literary inscription and helped allay any debate about whether the map was able to count as fact. Additionally, the two maps, as heterotopic, contested versions of space, work with and against each other to make their knowledge claims. As Turnbull says, for Latour, power is "the consequence of association," of the ability of words and images to make connections that mobilize and "muster allies on the spot—allies, that is, in the struggle over what is to count as a fact" (55). The claims made by the NRDC and NMFS are indeed claims rooted in the struggle over what gets to count as a fact. And even though the NRDC's map was limited in its ability to explicitly show evidence of an "alternative biogeographical scheme" that could be applied specifically for marine mammals, the map was still able to "raise serious questions on the merits" of the NMFS's claims (NRDC V. Evans, 2002a 11). Thus, the NRDC was able to gain an ally in the Court, at least in terms of the Court's agreement that the NMFS misrepresented the specified geographic areas involved in LFA deployment.

Lastly, and perhaps most important, these maps worked beyond the fact of their relative successes in the courtroom. The NRDC's map, while exhibiting certain limitations in the extent of the claims it could reasonably make, ultimately impacted the marine mammal bodies residing in the North Pacific Ocean, protecting them from the threat of LFA sonar. What this chapter has also shown is that the affected bodies need not be physically present in the setting of the debate and that visual-material rhetorical artifacts can work to advocate for bodies that might not otherwise be able to advocate for themselves. In other words, the map can function as an advocacy tool that helps protect those who cannot convey firsthand the impact of the debate on their bodies, but who shoulder the consequences of the debate nonetheless.

Today, the debate over the use of sonar in military training exercises continues. In November 2008, the issue went to the Supreme Court, which ruled to lift "judicial restrictions on submarine training exercises off the coast of Southern California that may harm marine mammals. In balancing military preparedness against environmental concerns, the majority came down solidly on the side of national security" (Liptak). As reported by *The New York Times:* "For the environmental groups that sought to limit the exercises, Chief Justice Roberts wrote, 'the most serious possible injury would be harm to an unknown number of marine mammals that they study and observe.' By contrast, he continued, 'forcing the Navy to deploy an inadequately trained antisubmarine force jeopardizes the safety of the fleet'" (Liptak). This sort of dichotomized rationale situates marine mammal safety as incompatible with the goals of national security, again perpetuating the nature/society, human/nonhuman binaries that foreclose possibilities of more embodied ways of knowing. Such binary relationships are easily perpetuated through the visual-material rhetoric of the map. We may recall that, for Brooke, the idea of a "posthuman rhetoric, as a return to embodied information, involves a revaluing of partiality. A posthuman rhetoric would allow us to turn our backs on omniscience and the humanist values of mastery and control that derive from the will to knowledge" (791). If we also understand the posthuman as concerned with broader issues of nonhuman ways of knowing, or as Harris and Hazen put it, a "more-than-human perspective" (50), then a continued visual-material rhetorics of the posthuman or more-than-human would indeed be attentive to how visual-material artifacts function in rhetorical contexts such as the debate over sonar use and marine mammal protection, or situations in which mapping practices may be linked to "hierarchies of values" that "privilege certain social groups" or species over others (Harris and Hazen 55). It is here that we can also begin to see connections between visual-material rhetorics and the burgeoning fields of animal studies and human-animal relations, which are often concerned with questions about the ways in which nonhuman animals are understood in relation to humans, the characteristics projected on to nonhuman animals, or the contexts in which specific species are deemed valuable or important. A visual-material rhetorics of the more-than-human could advocate for marginalized, nonhuman animal bodies and would perhaps begin to work against these binaries and hierarchical relationships, arguing instead for what Haraway has termed "embodied communication" (26). The idea of embodied communication is founded on the premise that we are "multiple beings in relationship" and may be defined as "communication about relationship, the relationship itself, and the means of reshaping relationship and so its enacters" (72; 26). Not incompatible with this thinking, Harris and Hazen note that "animals can be regarded as a 'marginal 'social' group' that is 'subjected to all manner of socio-spatial inclusions and exclusions'"; with this issue in mind, they ask how attention "to the spatiality and temporality of map production, uses, and engagements affect[s] the condition and effects of particular maps and mapping practices" (55). What this chapter reveals is that the maps involved in this debate, the NRDC's map in particular, played an integral role in environmental decision-making that resulted, at least temporarily, in the inclusion and protection of marine mammal lives by remaining mindful of contexts related to the spatiality and temporality of the rhetorical situation.

As we have seen demonstrated in this chapter, when we understand visual-material artifacts as rhetorical and analyze them from the vantage point of their audience, purpose, and social context, all of which Blair and Foucault's theories help accomplish, relationships between artifacts are illuminated that may otherwise remain less salient. This chapter also reveals the limitations of the work that visual-material artifacts can accomplish, and Foucault and Blair's theories are helpful in highlighting the ways in which the map functions as but one component of the rhetorical situation. Finally, the portrayal of biogeographic regions, the choice of map projection given the context, and the use of conventional signs and symbols all greatly contributed to the NRDC map's material force beyond its initial goal fulfillment. Again, this is not to say, however, that the NRDC's map fully resolved the immediate situation, that the broader issue in which the two maps were immersed was resolved in a tidy package, or that either of the maps truly engaged in the sort of embodied communication to the extent called for by Haraway. Nonetheless, the ability of visual-material rhetorical artifacts to help advocate for the nonhuman animal body is clear, and a visual-material rhetorical approach that focuses on the nonhuman animal body, or the "more-than-human perspective" provides a valuable and sustainable project of inquiry moving forward and should not be set aside with the concluding of this chapter; rather, it is my hope that these discussions may serve as catalyst for further explorations in visual-material rhetoric studies.

6 Locating Visual-Material Rhetorics

To understand visual-material artifacts as rhetorical means exploring not only their individual graphical, physical, or textual components but also their combined, broader consequences on the rhetorical situation and the bodies implicated within those settings. It requires examining not only the rhetorical responses to visual-material artifacts but how those responses are shaped by the presences of objects such as the commemorative sculpture, the map, or the GPS. The notion of a text's consequence, says Blair, is seldom the focus of our studies and is "frequently understood narrowly as 'success' or goal fulfillment" (21). While we may understand and acknowledge that rhetoric is not solely about production, that "it has consequences that exceed goal fulfillment," Blair feels that we must pay greater attention to these areas nonetheless. By devoting greater attention to a text's materiality and its impact on the body, we may gain some ground in better understanding rhetoric's "material force" past the mode of production (22).

To shift our analytical focus from the circumstances surrounding a text's production, symbolic value, and relative successes to its broader consequences requires that we work against some of our traditional understandings of a text's rhetorical import. Admittedly, this is sometimes easier said than done. This potential for viewing the rhetorical situation with the blinders of immediacy should then serve as a reminder of how much more there is to be gained from a perspective that, while of course not dismissive of a text's symbolicity, emphasizes to a greater extent materiality and the value of an approach attentive to embodied knowledge. Similar to understanding visual rhetoric as a mode of inquiry, a visual-material rhetorical approach has implications for rhetorical analysis that extend beyond the immediate object of study to its broader consequences in the rhetorical situation. The explorations undertaken in this book then help to illustrate the understandings that become possible when conceptualizing visual-material rhetorics not as an immediate product or outcome but as a sustained mode of inquiry. In the discussion that follows, I return to some of the earlier examples in the book, considering not only what it means to understand visual rhetoric as a project of inquiry but also how our understandings of visual rhetoric might shift if we considered such projects more so from the vantage point of their material dimensions. I discuss the implications of visual-material rhetorics for rhetorical analysis and for understanding rhetoric as advocacy work; I consider future directions for research in visual-material rhetorics, and implications for undergraduate and graduate pedagogy.

THE IMPLICATIONS OF VISUAL-MATERIAL RHETORICS

Finnegan's analysis of image vernaculars and the early photo of Abraham Lincoln, as described in the introduction, aptly demonstrates how visual rhetoric operates as a mode of inquiry. First, by understanding image vernaculars as enthymematic, she uses the tools of rhetorical theory to explore the artifacts of visual rhetoric and understand them as such. Conversely, she brings rhetorical theory into play in a unique way that responds to the challenges brought forth by visual artifacts. Next, by understanding readers' responses to the photo reproduction in *McClure's* as informed not only by myths about Lincoln himself but also by their social knowledge of photography and nineteenth century understandings of the discourses of physiognomy and phrenology, she is able to uncover the power and knowledge dynamics at play around how the photo makes meaning; that is, how readers come to understand it as presenting evidence of Lincoln's moral character. Finally, she embraces the interplay between the image and its surrounding textual artifacts (letters written by readers in response to the photo) as a new site for knowledge-making that draws upon both the practices of visuality and the knowledge of the tools of rhetoric to open up rather than close off interpretive possibilities.

Finnegan's analysis, however, does more than just serve as a viable project of visual rhetoric. When she uses the tools of rhetoric to explore the impact of photography on the practices of knowledge-making about the body, she is in many ways focusing on the material dimensions of visual rhetorical analysis. By exploring the connections between nineteenth century portraits and their assumed linkage to what they could reflect about human character, Finnegan implicitly shifts the emphasis of visual rhetoric to a focus on its material dimensions. That is, to understand responses to the Lincoln photo, she looks at the connections between the emergence of photography, the "discourses of morality," and the subsequent use of the photograph not only "for the purposes of criminal identification and classification" but also in the everyday "'practice of reading faces'" ("Recognizing Lincoln" 68). She not only links nineteenth-century discourses about physiognomy to how viewers of the Lincoln photo were able to conclude that "in Lincoln's face may be found the key to his character," but she goes further, moving beyond rhetoric's immediate effects, to acknowledge the broader rhetorical consequences of connecting a "hermeneutic of the face" to the potential "for a full-blown discourse of eugenics" (69–70). Within Finnegan's analysis of image vernaculars and the Lincoln photo, then, we not only see the criteria for a project of visual rhetoric fulfilled but we can begin to understand some of what it means for scholars to be attentive to the material dimensions of visual rhetorical analysis.

Likewise, Hariman and Lucaites also broach the material dimensions of visual rhetoric when they focus on the palpable physicality inherent in the Iwo Jima photo and the ways in which the photo acts on viewers. They perceive the actions of the Marines' bodies in raising the flag pole as making tangible the otherwise abstract concept of national identity: "The Iwo Jima image is more than another instance of nationalism because it so effectively grounds abstract national identity in embodied social performance" (100). Within the still life of the visual artifact, the Marines engage with the flag in such a way that creates an active relationship between their bodies, the landscape, and the object of the flag. Viewers can then imagine a performance unfold before their eyes, thus "giving the abstract nation living embodiment while giving collective labor the transcendental status of nation building" (100). To consider the material dimensions of visual texts and the ways in which photographs, for example, may be understood as "embodied social performance" is to begin to build bridges between visual and material rhetorics.

The discussion of photo 22727 in chapter one makes similar connections between visual and material rhetorics, though to different ends. Understanding maps as "the spatial embodiment of knowledge" and as encouraging new ways of knowing the world allows for an understanding of visual artifacts that is also active and engaged (Cosgrove, "Mapping Meaning" 1). Understanding photo 22727 in light of its status as iconic photo and in terms of the discourses sparked through its interpretation allows us to view the photo as a rhetorical artifact. To situate the photo within the social and political contexts of the emerging environmental movement allows us to understand the dynamics of power and knowledge reflected and perpetuated through the image and its subsequent appropriations over time. Finally, to understand the photo as both an iconic photograph and as an object of cartographic practice allows us to complicate how we define the work of visual artifacts and the rhetorical work they accomplish in the world by prompting us to shift our analytical emphasis toward a greater focus on space, place, and the body.

An analysis of photo 22727 thus serves as an important stepping stone, especially within the context of this book, that prompts us to further acknowledge an important aspect of visual rhetoric projects, which is the need to "(re)consider aspects of rhetorical theory" relative to the new challenges brought about through analyses of visual artifacts (Finnegan, "Rhetorical History" 198). That is, an analysis of photo 22727 as cartographic representation helps demonstrate that to better understand the consequences of the rhetorical work of visual-material artifacts in the world, we must consider rhetorical theory as also attentive to studies of space, place, and the body, an idea that the subsequent case chapters in this book work to further explicate. The question of how rhetorical theory can better respond to the interpretive challenges posed by visual-material artifacts, and thus the implications of visual-material rhetorics for rhetorical analysis, has underpinned the analytical work of this book. Through this analytical work, it becomes clear that a theory of visual-material rhetorics that couples Blair's theory of material rhetoric with Foucault's concept of heterotopias can allow for a more nuanced and embodied understanding of how visual-material artifacts and spaces function rhetorically in the world.

At the Lowell Mills National Historical Park (LMNHP), for example, the visual-material artifacts and the spaces in which they reside invite and encourage interactive, embodied experiences that contribute to the uniquely contextualized experiences of the moment. Again, a visual-material rhetorical approach does not eschew the value of symbolic meaning but focuses more so on how spaces are experienced by those who visit or inhabit them, the consequences of those experiences on the body, and how, through the practices of visuality and bodily experience, we come to better understand the rhetorical situation. The LMNHP relies on but also moves beyond symbolic meaning, insisting on a corporeal experience that demands a different way of engaging with these texts and the historic contexts in which are immersed. The rhetorical spaces and artifacts at the LMNHP have such a clearly visceral impact on the visitor that, once immersed in the spaces they occupy, it becomes more necessary and also interesting to physically negotiate these spaces and their consequences on the body than ponder the artists' intentions. The *Circular Fence,* for example, imposes on its visitors a non-linearity both in its textual inscription and physical design that is disorienting in itself, but even more so when set against the park's linearity. Visitors' experiences of the piece as at once aesthetically pleasing and unsettling can invoke the tension felt by the Mill Girls as they worked in a crisis heterotopia that was presented to the outside world as a tidy heterotopia of compensation. The *Path Markers* also invoke dissonance, in that they weave into the park a consistent thread that creates both continuity and destabilization. The *Seating Circle* too creates a feeling of discontinuity that may be read as performing the crisis heterotopia in which the Mill Girls lived and worked. The broken circle is suggestive of the disconnects between mill workers, superintendents, and mill owners; however, the message carved in granite within each section of the circle indicates that continuity is still possible and may be achieved. The sculpture acts on the body by inviting visitors inside it, to walk on its cobblestone floor, or to use the sculpture as a bench—to sit with its simultaneous discontinuity and hopefulness. The sculpture that perhaps best epitomizes the symbiotic relationship of visual-material rhetorics and heterotopic space is the *Fourteen Hour Clock.* As the most destabilizing element of *Industry, Not Servitude,* this piece, according to the park ranger, is the one that most resonates with visitors. The clock also incorporates textual inscription in its inclusion of the petition for the 10-hour workday. To read the petition alongside the clock requires a physically taxing engagement with the text—one that, like the rest of this sculpture and the others in Lucy Larcom Park, reflects and performs the crisis heterotopia of the early Lowell Mills, and in doing so, acts on visitors' bodies in ways that arguably make them more empathetic to the Mill Girls' experiences.

Sculptures One, Two, and Three in Boardinghouse Park serve primarily to demarcate the perimeter of Boardinghouse Park and function symbolically as a testament to Lowell's industrial heritage. In the sculptures' gesturing toward the hard work and cramped living quarters of mill workers generally and the Mill Girls specifically, they reflect the crisis heterotopia of the Lowell Mills. Additionally, the three sculptures emphasize the concert stage, which, built over the former site of two boardinghouses, maybe be understood as commemorating and celebrating the lives of the Mill Girls. The stage epitomizes the ideal heterochrony, in that it represents both the ongoing festival, as it is quite literally the site of the annual Lowell Folk Festival, as well as "the eternity of accumulating time," in its situatedness over the former boardinghouses (Foucault, "Of Other Spaces" 26). Finally, the juxtaposition of the Agent's House with the former boardinghouse next to the park allows visitors to understand the spatial relationships that contributed to the crisis heterotopia in which the Mill Girls lived.

The work that most literally depicts the bodies of the Mill Girls is *Homage to Women*. Composed of granite and bronze, the sculpture's durability performs the persistence and endurance required of the Mill Girls' circumstance. Because the sculpture is made of bronze, the expressions on the women's faces do not photograph easily, and are therefore not as fully discernable as they are in-person. To see the sculpture in-person is therefore a much more powerful experience than to view its reproduction on the park's web site, or even in the digital photographs reproduced in chapter three. *Homage to Women* requires a physical engagement with the piece in order to fully discern its features. The visitor sees these women's faces as appearing haunted, concerned, and pained; at least four of them appear to be crowding one another. Their expressions, gestures, and movements constitute a powerful visual-material rhetoric of the Mill Girls' lives. The sculpture's material existence deems their history attention-worthy and puts them on the map of Lowell's contemporary historical landscape.

Indeed, many facets of the LMNHP may be understood as reflecting the crisis heterotopias and heterotopias of compensation of the early mills. Perhaps one consequence of visitors' corporeal experiences of these heterotopic spaces is a greater understanding of and empathy for the lives and daily experiences of the Mill Girls. Thus, a visual-material rhetorical approach informed by the theories of Blair and Foucault can function in the service of advocacy, such that audiences may engage with greater empathy in the lives and struggles of underrepresented groups or historically tangential voices, such as, in this case, the lives of the Mill Girls.

Visual-material rhetoric can also help illuminate the embodied knowledge that is fostered through the relationship between rhetorical artifacts and the technologically mediated, posthuman body. GPS users, for example, work with and against this navigational technology to make purposeful, informed decisions. Through its physicality and multimodal cues, the GPS engages the user and invites an audience—in some cases even a dialogue—that, once again, requires the user to remain mindful of both symbolic and material environments. Users interact simultaneously with the GPS's symbolic, visual cues and culturally-bound symbols, and the physical terrain of the material world represented by those symbolic displays. As a heterotopic artifact, the GPS produces digital, symbolic, cartographic texts that not only bear relation to the surrounding environment but might also challenge or compete with users' knowledge of their surroundings. Such multimodal, interactive engagement with the GPS participates in the construction of geographic knowledge that then informs users' decision-making. Agency becomes an interactive, mediated process that is co-constructed by the GPS and its user and is constituted by the embodied, geographic decision-making derived through these relationships. Here we see visual-material rhetorics function in the service of advocacy as they help foster acts of purposeful decision-making.

Such decision-making and knowledge-making practices are borne out of rhetorical interactions that involve supplementing the GPS and using the GPS supplementally; correcting the GPS and being corrected by it; programming and reproducing routes and challenging those reproductions; engaging in digital memory practices by storing and retrieving addresses and other geographical information; negotiating and challenging its routes through subtle subversions; and engaging with its symbolic representations while remaining attentive to actual physical surroundings. On the one hand, these rhetorical interactions have consequences relative to immediate goal fulfillment that, presumably, involve learning about or navigating to a specific destination. On the other hand, however, as a visual-material rhetoric helps uncover, the rhetorical actions associated with GPS use have broader consequences for the mediated, posthuman body, beyond those of immediate goal fulfillment. The rhetorical consequences of GPS use surpass the immediate goals of destination arrival to affect the broader social and interpersonal relationship dynamics that may be linked in different ways to those navigational goals.

In the face of constantly shifting political, social, and technological contexts, it becomes even more useful to employ a rhetorically analytical approach that accounts for purposefulness beyond immediacy. In July 2009, for example, I was in the midst of considering the visual-material rhetorical impact of the GPS device, when an article titled "Sending GPS Devices the Way of the Tape Deck?" appeared in *The New York Times*. Touting the smartphone as the "Swiss Army knife of the digital age," author Jenna Wortham questioned whether it would soon displace the standalone GPS device ("Sending"). Interestingly, one GPS user interviewed in chapter four made a comparison between these two technologies when likening the sense of security provided by the GPS to that of traveling with a cell phone: "[I]t's sort of like having your cell phone on you [. . . W]hen I traveled without a phone, there are things that I wouldn't have done, that I'm not worried about doing now, because I know that I'm a phone call away from talking to somebody [. . . B]ut the GPS, I don't know, to me, it's almost as essential as a cell phone" (OFT2). GPS users are beginning to merge the utility of these two devices in ways that will perpetuate new modes of embodied knowledge-making. Technological shifts such as this should serve as further evidence that we require a methodological and analytical framework to account for such constantly changing modes of knowledge-making and technological mediation. Moreover, the idea that visual-material rhetorics can account for shifting epistemological contexts need not remain constrained solely to its implications for human bodies.

Visual-material rhetorics can illuminate competing knowledge claims about contested space in ways that can shape environmental policy-making, as we saw in chapter five. While mindful of the symbolicity and purposefulness of the NMFS and NRDC maps, a visual-material rhetorical analysis illuminated not only their immediate differences but also showed that a comparison of their graphical features and cartographic conventions conveyed only part of the story, and that the knowledge claims implicit in these maps worked beyond the fact of their relative successes in the courtroom. Again, while it is useful to explore the conditions surrounding the text's immediate goals, it becomes even more important to understand the text's larger consequences beyond initial goal fulfillment in the most immediate and visible aspects of the rhetorical situation. While the claims made through each of the two maps certainly represent claims rooted "in the struggle over what gets to count as a fact," most critical are the consequences of this struggle (Turnbull 55). By first understanding each map as a representation of heterotopic space, and subsequently by juxtaposing the two texts, it becomes clear that each makes subtly different claims to knowledge, and that together, the maps constitute versions of contested, rhetorical space. Foucault's theory of heterotopias becomes helpful because it allows for the juxtaposition of two incompatible versions of a single territory, to better emphasize the differing spatial relationships and to understand those differences as rhetorical. Foucault's theory of heterotopias also illuminates the ways in which these maps may be understood as linked to slices in time, or heterochronies, during specific cultural moments, out of which then arose a perceived need for social and environmental change. This chapter also showed how maps can act on behalf of bodies that cannot be physically present in the setting of the debate. Working in the positive, maps may be understood as rhetorical advocacy tools that can represent groups that might not otherwise have representation in the debate.

Conversely, as chapter five also showed, genres such as conservation maps and environmental policy maps have the potential to perpetuate and sustain the binaries that reinforce hierarchical relationships between humans and nonhuman animals. A visual-material rhetorical approach might view an analysis of these artifacts as an opportunity to interrogate these divides, understanding a focus on human action as always already implicated in and bearing consequence on nonhuman bodies. Maps then function as powerful artifacts of visual-material rhetoric; without the analytical lens afforded by a visual-material rhetorical approach, artifacts such as these and the contexts in which they are immersed might be viewed as significant only in terms of their immediate goals and outcomes.

Again, the contexts in which visual-material artifacts are situated are always shifting and subject to political, social, and technological changes. While the two maps involved in this debate each functioned within the context of a specific debate over the deployment of LFA sonar in the North Pacific Ocean, the debate over sonar use in military training exercises continues today and even made its way to the Supreme Court in November 2008. The 2008 case again saw the Natural Resources Defense Council at loggerheads with the United States Navy; as Charles Siebert of *The New York Times* describes, the

NRDC had already won two previous "landmark victories in the district and appellate courts of California, which ruled to heavily restrict the Navy's use of sonar devices in its training exercises" ("Watching Whales"). This time, however, the Supreme Court "overturned parts of the lower-court rulings [. . .] in a 6–3 decision that was widely viewed as a setback for the environmental movement" ("Watching Whales"). Interestingly, Siebert describes the reactions of environmental groups as focused not on the immediate outcome of the case, but rather on the larger consequences of the debate, when he writes that "the majority's verdict somehow seemed incidental to the greater, tacit victory for environmentalists of having gotten the nation's highest court to even consider the well-being of whales in the context of a debate about national security, something that would have been unthinkable not so very long ago" ("Watching Whales").

That environmental advocacy groups like the NRDC recognize the value of remaining focused on the larger consequences of debates over LFA use and marine mammal protection, even in the face of a Supreme Court loss, should serve to signal the importance of an increased scholarly focus on the broader implications of the rhetorical situation—implications that are quite often illuminated through a visual-material rhetorical approach. As Blair states, "rhetoric occurs in a pedestrian world and exerts its most important consequences in the realm of human affairs" (51). Thus, she says, we must remain attentive to the consequences of rhetorical action in the social world (51). A visual-material rhetoric that is mindful of rhetorical action in the social world must attend not only to immediate goal fulfillment but also to the broader consequences of those actions on all bodies affected by them. I contend, then, that a concern with the bodies affected by the consequences of rhetorical action should extend beyond a concern with human or posthuman bodies to that of the nonhuman animal body as well, broaching what Harris and Hazen call the "more-than-human perspective" (50). Such understanding is not incompatible with Brooke's call for "a return to embodied information" and "a revaluing of partiality" (791). That is, if we also understand the posthuman as concerned with broader issues of nonhuman ways of knowing, then visual-material rhetorics would indeed be attentive to the rhetorical contexts affecting both human and nonhuman bodies. The approach modeled in this book and the analyses taken up through its modes of inquiry provide a means for better understanding how these ways of knowing are made possible. But what next, then, for visual-material rhetorics?

FUTURE DIRECTIONS

To further conceptualize visual-material rhetorics and its implications for rhetorical analysis means considering not only future work in the field but also its continued discussion within the classroom and through research. I argue in this book that to conceptualize visual-material rhetoric means thinking not only in terms of Blair's approach to material rhetoric and its focus on consequence, but also in terms that understand space as heterotopic and rhetorical. Future work might further explore the implications of visual-material rhetorics for technologically-mediated environments such as virtual museums, online games (we might consider the strong focus on gaming that emerged during the Computers and Composition 2010 conference), or other multimodal sites in which knowledge is spatially embodied, considering such spaces as rhetorical, geographic, and heterotopic. Heterotopias allow us to characterize and contextualize space in ways that illuminate its heterogeneous, selective, shifting, and sometimes contested nature; such understandings of space then help illuminate its rhetorical power. Understanding space not only as rhetorical but also as informed by perspectives aligned with critical cartographies and human and feminist geographies then allows us to discern the multiple consequences of space on multiple kinds of bodies. The questions invited by an interpretive framework that combines Blair's notion of material rhetorics with Foucault's heterotopology could allow those who study visual artifacts such as photographs, advertising images, or other more traditionally visual works of art to consider the impacts of these genres on the practices that contribute to embodied knowledge-making, or the material dimensions of visual rhetorical analysis. I have also expressed throughout this book a perceived need for visual-material rhetorics to function in the service of advocacy. To my mind, that is, underpinning visual-material rhetorical analysis is the idea that rhetorical artifacts and spaces can help advocate for tangential or underrepresented groups, or help foster a sense of agency in those who interact with such artifacts or within rhetorical spaces.

Understanding Visual-Material Rhetorics as Advocacy Work

As the cases in this book help to demonstrate, advocacy work can play a large role in visual-material rhetorics and often manifests in different ways across artifacts and analyses. Thus, future work in the field could also entail comprehensive analyses of the visual-material, multimodal, and textual artifacts involved in issues of rhetoric and environmental advocacy. It could, for instance, consider how visual-material, multimodal, and other textual artifacts and genres (such as parks, environmental centers, GIS maps, and environmental impact statements) function together, and rhetorically, within a broader activity system to advocate for species preservation and biodiversity within particular ecosystems. The notion of activity theory, as related to genre studies within the fields of rhetoric and composition and professional writing could complement such work because it is understood as need-based and as taking place within a system in which groups make use of varied tools or artifacts to work toward a particular goal or outcome (see Propen and Schuster, 2009). Moreover, such tools or artifacts do not merely help make connections between groups and the objects with which these groups might work; rather, such artifacts, through their acts of mediation, qualitatively change the types of activities in which subjects engage (see Spinuzzi, 1996; 2003). Given these ideas, future work might consider how rhetorical artifacts function as mediating devices within various, multimodal acts of ecosystem preservation or conservation mapping, and how they help accomplish meaningful advocacy work through the combined activities and interpretations of biologists, cartographers, conservation groups, concerned citizens, and other related parties. Such work could easily entail partnerships with local or regional organizations.

Visual-Material Rhetorics: Implications for Research and Pedagogy

Working with groups outside of academe can provide a unique and helpful vantage point and can also be highly fulfilling. The idea of what it means to have a research partner may be conceived of in different ways. It may involve, for example, as it did in my work with the Lowell Mills National Historical Park, a combination of interviews and subsequent communications or follow-up discussions with park officials over time; in other words, a relationship is established over time that is not only fulfilling but can also better inform scholarship. Partnerships can also involve more integrated relationships, in which volunteer work with an organization may spark further research related to or involving the organization or its surrounding discourses.

Partnering with an organization can have implications not only for scholarship in the field but for graduate pedagogy as well. As Mary Lay Schuster and I have argued elsewhere, "graduate courses need to address [. . .] more nuanced relationships within the public arena, [and] should also subsequently encourage firsthand experiences within those arenas whenever possible" ("Making Academic" 323). These firsthand experiences then help contextualize theory and work against the tendency toward academic insularity: "By encouraging partnerships with organizations that deal with various forms of public policy, whether it be in the legal or nonprofit arena, for example, students get to witness policy in action versus or in addition to policy analyzed in the text" (323). Research partnerships can also foster new understandings of policy and advocacy within the organizations with whom we partner, as well (see Lindeman, 2007).

To partner with an organization or even to make connections with research participants in the context of ethnographic work or interviews also provides new opportunities for sharing the results of our work with new audiences; moreover, "[s]haring research results with participants and inviting their participation or feedback is also an important component of activist research" (Propen and Schuster, "Making Academic" 324; see also Barton, 2004; Faber, 1998; Grabill, 2000; Schuster, 2006; and Waddell, 1996). Thus, graduate coursework in research methods in fields such as rhetoric and composition, technical communication, communication studies, human geography, and critical cartographies could all "benefit by addressing the complexities of sharing research results with research participants or the public" (Propen and Schuster, "Making Academic" 324).

Of course, as teachers of rhetoric and writing, our research invariably informs our pedagogy, and so it makes sense that ideas about visual-material rhetorics would then surface in our undergraduate classrooms as well. Like scholarship in visual rhetoric, the study of the visual in the composition classroom has historically taken a "secondary position" to the study of the text (Richards and David 3). With increased scholarly attention to visual ways of knowing, however, and alongside the emergence of new media and new generations of college students who have grown up with technology as an ever-present given, modes of writing and communication now include the visual "'not as

attendant to the verbal but as complex communication intricately related to the world around them'" (3). Rather than understand attention
to the visual as taking time away from writing, we may understand the
two modes as interrelated; thus, as Richards and David note, reading
and writing time "need not be affected by our including visual texts
among the verbal texts we assign students" (5). In fact, students who
have grown up in a technologically-mediated society often implicitly
expect that the visual and multimodal are part of everyday communication: "Students are much more likely than we are to be immersed in
visual culture and to feel comfortable talking and writing about what
they see" (5). Thus, to integrate analysis of the visual into writing assignments can "intensify student engagement with assignments" (5).
Along with integration of the visual into writing assignments, teachers
of rhetoric and composition would also do well to understand writing
as necessarily attentive to the multimodal; as Richards and David then
suggest, "we must acquire to the best of our abilities the skills needed
to interpret and to create images and sounds and integrate them elec-
tronically with discourse" (4).

For those teachers who may feel uneasy at the thought of integrating
visual, material, or digital writing assignments into their undergraduate
writing courses, I offer a couple of suggestions. First, I maintain that
John Berger's essay, "Ways of Seeing," can be a fruitful and accessible
point of entry into discussions of the impact of the visual within society. Again, Berger focuses largely on the ways in which reproductions
and appropriations have made famous works of art accessible to the
public, but because those reproductions tend to manifest mostly in advertising and "[f]ashion and beauty photography," they perpetuate representations of "wealth and luxury" that not only "feed capitalism" but
also create a sort of mystification that distances viewers from the work's
original context and meaning (Richards and David 7). Based on the
ideas in this essay it becomes easy and interesting to generate writing assignments that ask students, for example, to explore their own assumptions about images. (If using *Ways of Reading,* for example, the essay is
nicely accompanied by writing prompts and assignments.) Possibilities
exist, for example, for asking students to visit a local museum and analyze a painting or work of art to which they are especially drawn. Prior
to reading any informational plaques associated with the piece, they
should spend some time free-writing about the painting, based only
on their own thoughts about what they see or the story they imagine it

to tell. After crafting this narrative, students might then research the painting's original meaning and context and write a follow-up analysis that compares their new knowledge with their initial ideas and assumptions.

Countless possibilities also exist for integrating multimodal writing assignments into the composition classroom, and ones related to understandings of place, at that. For example, within the context of discussions about digital writing, students might consider how web-based mapping tools like MapQuest and Google Maps have impacted how we acquire and interpret directions (see Carr, 2008; Flynn, 2009; Weeks, 2010). Students can use tools like Google Maps to create their own personal narratives about place, focused on their daily routine, a map of their childhood, or a map of past or upcoming travels. Google Maps allows its users to create such maps, make them public, keep them private, or send the map as a URL to an individual person, making it easy for students to submit their maps as assignments via email or through course management tools. Students also enjoy sharing their maps with their classmates and often make new social connections within the classroom, as a result.

Finally, L.J. Nicoletti's writing assignment related to memorial creation helps us understand the ways in which ideas about material rhetoric and writing about space can be made accessible for undergraduate students and subsequently serve as powerful tools for critical thinking. Again, Nicoletti's students analyzed "a site that was of special interest to them and then designed and wrote a text justifying the creation of a new memorial" (Richards and David 24). In the case of Nicoletti's assignment, students were also, as a result, able to better process the events of September 11, 2001. Projects of material rhetoric are also compatible with classroom research related to local ethnographies and analyses of information spaces, and are well-suited to help contextualize the study of urban histories. For an online material rhetorics of the "poetry and prose" embedded within the urban monuments of Boston's Southwest Corridor, for example, see "Writing on the Line" (Sullivan et al., 2008).

The integration of visual-material rhetorics in the undergraduate classroom, then, can help emphasize the notion of writing as social, political, material, and spatial. By learning to recognize the persuasive components of visual-material and multimodal texts, students may, by extension, be more well-equipped "to articulate what constitutes written persuasion or even argument" (Richards and David 5). As a result, students may come to more explicitly understand the interplay between "visual and written cultures," and may develop a clearer understanding of the digital and spatial components of writing (5). Moreover, by continuing to integrate visual, spatial, and multimodal ways of knowing into undergraduate writing pedagogies, we not only share with students our enthusiasm for our work but we invariably develop new ways of conceptualizing and communicating theoretical constructs and methodological approaches to studying visual-material rhetorics.

As the analyses in this book have shown, scholarship in visual-material rhetorics easily invites a variety of approaches to uncovering the various rhetorical dimensions of visual-material artifacts. In this book, for example, I have primarily adhered to an interpretive framework that combines the theories of Blair and Foucault, though my methods of data collection have varied given the scope and subjects of my analyses. A visual-material rhetorical approach easily invites and even anticipates a mixed methods approach to data collection, as well as the use of unique interpretive frameworks. The tools of data collection may involve, when relevant and necessary: interviews, observation, ethnographic work, internet research, historical and archival research, photography, the creation of field notes, rhetorical analysis, and working both directly and indirectly with groups or organizations related to the objects of study. A mixed-methods approach, however, adds an extra layer of complexity to research projects. For graduate students who must clearly articulate their method of data collection in any given research project, a mixed-methods approach undoubtedly leaves more to account for in the way of description and justification of approach. To combine archival research with observations and interviewing, however, often creates room for richer and more nuanced description and analysis. While one method may ultimately become ancillary to the other, such combinations may still help provide a more holistic picture of the object of study. In addition, modes of data collection inevitably shape or influence interpretation to some extent, and so it is necessary to consider the benefits and limitations of different approaches. Theory itself can also serve as an additional interpretive tool, and is arguably more valuable when contextualized and applied than when presented or extended on its own.

To use theory as an interpretive tool, as, for example, Finnegan arguably does in her use of image vernaculars, allows theory to function as an interpretive tool first and as a contribution to theory second. In other words, as Mary Lay Schuster and I have expressed, it is useful for "graduate students have the opportunity to experience the application of theory as interpretive tools. Students often feel they must extend theory to succeed in their courses or dissertations, but the case study approach [. . .] demands that theory provide tools first to interpret and then to contribute to theory ("Making Academic" 323–324). To view theory as an interpretive tool not only allows for richer and more nuanced description and analysis but can also make research and writing more accessible to broader audiences of students, practitioners, and scholars alike.

In this book, I have sought to illuminate a more holistic understanding of visual rhetorics by better accounting for its material dimensions along the spectrum of what I have then viewed to be a visual-material rhetorical approach. To view visual-material rhetorics in this way first means understanding visual rhetoric as concerned with and receptive to studies of space, the body, and materiality. Such an understanding is valuable not only because it better accounts for the notion of space as rhetorical, but also because it more explicitly accounts for embodiment as an important component of visual rhetorical understanding. As Dickinson and Maugh remind us, to understand visual, multimodal representations and physical structures as concerned with more than their immediately apparent features—to understand them also as embodied—allows us "to locate our bodies in relation to other bodies in the world" (272). To understand visual rhetoric as concerned with spatiality and embodied knowledge, then, is to more fully conceptualize the broader implications and consequences of the rhetorical work of visual-material artifacts in the world, thus enabling the more inclusive and sustainable project of rhetorical inquiry that is constituted through a visual-material rhetorics. Together, the analyses taken up in this book illustrate the ways of knowing that become possible when conceptualizing a visual-material rhetoric that emphasizes the value of embodied knowledge and also understands rhetorical work as able to function in the service of advocacy. It is my belief that visual-material rhetorics, when understood in this light, can constitute an inclusive and sustainable project of inquiry moving forward, and it is my hope that this book can serve as a starting point for the continuation of such work.

Appendix A: Interview Questions, Chapter Four

1. Can you describe your background or familiarity with the GPS?
2. Can you recall and describe your initial impressions when first using the GPS?
3. Describe a typical scenario using your GPS. You do not need to reveal any personal details.
4. Are there particular features of the GPS that stood out to you while using it?
5. What do you like about the GPS?
6. What do you dislike about the GPS?
7. Did the GPS change your driving habits, driving routine, or decision-making at all? If so, can you describe this in more detail?
8. Is there anything else you would like to share?

Appendix B: Coding Categories and Subcategories Derived from Interviews in Chapter Four

Category	Subcategories
Descriptions of the surrounding environment and places	References to winding roads, confusing roads, back roads, poorly marked roads, strange or new places.
	References to "unfamiliar" places; using the GPS in unfamiliar places.
	References to common places such as grocery stores, gas stations, restaurants, and others.
Use of the GPS, as opposed to MapQuest, Google Maps, or other web-based mapping tools	No longer require printed directions.
	Supplementing with or checking directions against MapQuest, Google maps, etc.
	Comfort of paper-based directions.
Impact of GPS on knowledge of surroundings	No longer afraid of driving to new places; takes more risks; explores more freely.
	References to sense of direction.
	References to landmarks.
	Memorized or known routes.
	Being in the city.
	Use for longer trips.

Category	Subcategories
	Arriving at a destination.
Local knowledge	GPS not used in familiar places; GPS doesn't know local roads as well as the driver.
	GPS knows local roads or back roads.
	Driver learns short cuts through GPS use.
Fear of the GPS	Fear of the GPS when it was new.
	Getting comfortable with the GPS.
Getting lost	GPS gives inaccurate directions.
	No longer afraid of getting lost.
GPS as distracting or dangerous	Distracts driver from the road.
	References to getting in accidents.
	Programming while driving, viewed as either helpful or dangerous.
	Pulling over to use the GPS.
Programming the GPS or changing settings	Saving routes.
	Pushing the "home" button.
	Entering an address.
GPS as helpful	GPS makes driver feel more comfortable or safer.
	GPS helps driver pay attention.
	GPS saves time in traffic.
References to features	Touch screen.
	"Avoid" feature.
	Displays phone numbers of places.
	Other features.
Physicality of the GPS	Forgetting or forgetting about the GPS.
	Keeping the GPS in the car.

Category	Subcategories
	Physical presence of the GPS.
	Deciding where to put the GPS.
	Physical interactions with the GPS.
Visual display	Size of screen.
	Watching the screen.
	Being able to see where to turn.
	GPS displays different perspectives or views.
	Other descriptions of visual display: icons, colored lines, or arrows.
Audio cues	Voice can be helpful.
	Voice can be "annoying."
	Attributing voice to need for dialogue.
	Voice settings, accents, pronunciations.
Communicating with the GPS: Driver talks to, consults with, or questions the GPS.	Driver feels in control.
	Dialogue or interaction with the GPS.
	Arguing with or resisting the GPS.
	Driver didn't necessarily know the best route after all; wishes they had listened to the GPS.
	Driver thinks they are smarter than the GPS.
	Naming the GPS.
	References to gender.
Communicating with the GPS: GPS "talks to" or acts on the driver.	GPS tells driver where to go.
	GPS "takes the driver" places.

Category	Subcategories
	The GPS doesn't know everything.
	The GPS "knows things."
	GPS communicates through silence.
	GPS takes driver a different way than they think is right.
	GPS allows driver to think less.
Negotiations with the GPS and power/control issues.	Deciding when to listen to the GPS or pay attention to it.
	Tricking or manipulating the GPS.
	Trusting the GPS.
	Giving up control to the GPS.
How GPS describes directions	Referring to a specific street name vs. "turn left."
	GPS tells driver how long before a turn.
	GPS doesn't provide enough advance notice before turns.
	GPS tells driver how long before arrival.
	Autocorrecting, recalculating, and u-turns as a good or neutral feature.
	Autocorrecting, recalculating, and u-turns as an annoyance.
Impact of GPS on personal life, social life, or relationships	Changes relationship with passenger.
	Impacts decision-making.
	Lifestyle changes.
	Takes extra time.
	Can get going very easily: "get in the car and go."

Category	Subcategories
Drawbacks to the GPS	Needing to download maps; references to upgrading.
	Expecting more from the GPS; Expectations were different or not met.
	Acknowledging imperfections.
	Losing or acquiring signal.
Having fun with the GPS	References to geocaching.
	References to playing with the GPS.
	Having fun with the GPS.
Other ideas about or descriptions of the GPS	
Subcategories	Media images of GPS.
	"Urban legends" about the GPS.

Notes

INTRODUCTION

1. Within the past ten to fifteen years, the field of rhetoric and composition has seen an increased interest in visual rhetoric and visual culture, accompanied by greater focus on and development of scholarship in the area. As Hariman and Lucaites put it, the subdiscipline has become a "veritable growth industry" in recent years (308). While it is impossible to provide a complete bibliography of relevant works here, I highlight below some useful reading for scholars new to visual rhetoric and visual culture.

Those looking to broaden their understanding of modes of visual analysis will undoubtedly appreciate the feminist geographer Gillian Rose's *Visual Methodologies* (2001), as well as Gunther Kress and Theo van Leeuwen's *Reading Images: The Grammar of Visual Design* (1996); Marguerite Helmers's *The Elements of Visual Analysis* (2005); and Marita Sturken and Lisa Cartwright's *Practices of Looking: An Introduction to Visual Culture* (2001). Charles Kostelnick and Michael Hassett's *Shaping Information: The Rhetoric of Visual Conventions* (2003) is also a helpful work for academics and practitioners alike, and charts nicely how visual conventions take hold in social groups and across genres of professional communication. Next, because studies related to visual communication, visual culture, and visual and material rhetoric convey such a diversity of analytical approaches, edited collections in the field abound. Those looking to line their shelves with a wonderful array of sources would do well to acquire Lester Olson, Cara Finnegan, and Diane Hope's *Visual Rhetoric: A Reader in Communication and American Culture* (2008); Charles Hill and Marguerite Helmers's *Defining Visual Rhetorics* (2004); and Carol David and Anne Richards's *Writing the Visual: A Practical Guide for Teachers of Composition and Communication* (2008), among others. These comprehensive collections on visual rhetoric also include essays on material rhetoric and convey a diversity of analytical approaches relevant to rhetoric, composition, and communication studies. Additionally, Sharon Crowley and Jack Selzer's *Rhetorical Bodies* (1999) and Lawrence Prelli's *Rhetorics of Display* (2006) include a more sustained focus on material rhetorics specifically. Collections such as Carolyn Handa's *Visual Rhetoric in a Digital World* (2004) and Mary Hocks and Michelle Kendrick's *Eloquent Images: Word and Image in the Age of New Media* (2003) focus on the intersections of visual rhetorics and digital media. Finally, as Hariman and Lucaites also point out, scholarly journals such as *Invisible Culture: An Electronic Journal for Visual Culture,* the *Journal of Visual Culture,* and the *Journal of Visual Literacy* all publish scholarship relevant to visual rhetorics and visual culture.

2. Of course, studies in visual rhetorics span genres such as political cartoons (see Edwards and Winkler, 2008); advertising images (see Hope, 2004); the fine arts (see Helmers, 2004); film (see Blakesley, 2004); medical and scientific illustrations (see Kostelnick, 2004); memorials (see Blair and Michel, 2008); electronic games (see Mullen, 2008); and others. My point here is not to under-define visual rhetoric's analytical purview but rather to chart an accessible point of entry for the new reader who might have limited familiarity with the subdiscipline. For a highly comprehensive discussion of visual rhetoric's relevance across genres and its reception within the humanities and in communication studies in particular, see Olson et al.'s "Visual Rhetoric in Communication: Continuing Questions and Contemporary Issues."

3. The enthymeme, as Finnegan describes, is an argument in which the premise or premises "are suppressed or assumed" ("Recognizing Lincoln" 63). The relevance of the enthymeme for visual rhetoric as a project of inquiry, she says, is in the fact of its contextual specificity and in its connection "to the everyday experiences of audiences." (63). That is, the audience makes an assumption about what they see or what they know based on their "tacit social knowledge" (63). She says, for example, that to refer to an image as *Photoshopped* requires the audience's prior knowledge of the software itself. The unstated premise then is context-specific and relies on social knowledge for its understanding. In situating image vernaculars as "enthymematic modes of reasoning," Finnegan understands them as "tacit topoi of argument that viewers employ creatively in specific rhetorical situations" (63).

4. For a more comprehensive and nuanced analysis of the Ground Zero photo, see Hill and Helmers (2004) or Hariman and Lucaites (2007).

5. In fact, it is visual rhetoric's ability to broach the discursivities of space that likely accounted for my initial gravitation toward this subdiscipline of rhetorical studies in the first place. That is, during my earlier training in both geography and English, I found myself frequently straddling the line between spatial and textual analytical modes. At the time, I did not understand that this line need not be a line at all—that it could instead function as a site of interplay and was something I had perhaps constructed to ease my indoctrination into these two disciplines. Eventually, through an introduction to the field of rhetoric and composition, and through industry work that spanned the fields of technical writing and related work with

digital mapping technologies, I became interested in the explanatory power and intersections of visual and material artifacts, their rhetorical function in everyday contexts, and their implications for knowledge-making.

CHAPTER ONE

1. For example, in response to a question posed at the 2003 Alliance of Rhetoric Societies (ARS) Conference on the status and future of rhetorical studies, David Blakesley was asked: "What is visual rhetoric, and what is *its* tradition?" In the position statement he prepared in response (excerpted here), Blakesley replies:

> To understand visual rhetoric better, we need to reanimate its tradition, and in doing so, reconsider our conception of rhetoric itself as primarily a verbal art. The differentiation of the verbal and visual runs deep not only in rhetoric but in mainstream epistemology. I propose that, rather than perpetuate this division of the verbal and visual, we now need to consider their common basis in perception. What we find is that even in everyday verbal expression (orally or in writing) there are profoundly complex visual components. We find also that there are verbal (and rhetorical) components in everyday acts of seeing. What insights does our rhetorical tradition provide on the nature of visual rhetoric? On the nature of seeing and its relationship to verbal and rhetorical processes? ("What is Visual Rhetoric?")

In a recent review of Hocks and Kendrick's edited collection, *Eloquent Images: Word and Image in the Age of New Media,* Laura Gurak and I noted that some of the contributors

> might have pressed more quickly past the English literature-as-high-culture/visual-as-low-culture polarity. [The] [e]ditors [. . .] in fact do a nice job making this move quite clearly in the collection's introduction, invoking the concept of hybrid [one that helps them account for multimodal literacies] and moving past the old discussions into what they (and we) believe should be the real focus of scholarship on visual rhetoric and literacy. (Gurak and Propen 108)

2. As Alan Gross described in his recent analysis of verbal-visual interaction in Lavoisier's final geological memoir, "arguments that incorporate the visual cannot be explained by any theory designed, like traditional rhetorical theory, solely for verbal analysis" (149). Rather, visual artifacts require their own methodological treatment that must account for the "differences between these two communicative modes" (149). Gross makes this distinction within the context of debates about whether a visual can make an argument "in the absence of words" (147). As he describes, this question has sparked

a good deal of debate. While the terrain of this particular debate focused primarily on the verbal and visual and does not explicitly account for other multimodalities or genres that rely on the interplay of text, image, space, materiality, or sound, some helpful distinctions have arisen. Keith Kenney, for example, has questioned whether a visual can make an argument if it cannot be understood as propositional form: "When rhetorical critics use the word *argument* they mean the presentation of premises followed by a conclusion" (58). Likewise, Randall Lake and Barbara Pickering noted that "[t]he problem of refutation in visual argumentation, then, occurs because even if pictures argue, they do not argue *propositions*" (80). Kenney then suggested that visuals can "persuade by argument when we have the ability to choose. Visuals also must: 1. provide reasons for choosing one way or another; 2. counter other arguments [. . .] and 3. cause us to change our beliefs or to act" (59). Kenney's criteria for when a visual may persuade seems to allow for persuasion without the visual's inclusion of propositional form. Gross also notes that while it is problematic to assert that images can be propositional, "*evidence* need not be propositional," and images such as photographs may certainly present evidence for a claim (148, emphasis added). Thus, we may infer from a picture certain relationships or claims, even though a wholly pictorial image such as a photograph may not include a proposition consisting "of terms and their predicates" (148). The relationships and claims that we may infer from a visual artifact may then provide us with a rationale that accounts for why we might be persuaded one way or another. If one idea has become clear through these discussions, it is that we cannot assume the verbal and visual to be mutually exclusive. While they constitute different modes of communication and representation, we must consider the terrain of their interplay through the exploration of various vantage points, disciplines, and interpretive frameworks.

3. Readers familiar with Plato's *Phaedrus,* for example, will find an irresistible and apt description of sense of place, or rather, out-of-placeness, articulated at the outset of the work, when Socrates finds Phaedrus about to embark on a stroll outside the city walls. This walk comes at the advice of Phaedrus's physician, Acumenus, who suggests that it is "more refreshing to walk along the country roads than the city streets" (Plato 227b). Phaedrus has invited Socrates to accompany him on the walk, where he will recount for Socrates, Lysias's speech: "You'll hear about it, if you are free to come along and listen" (227b). Socrates's great interest in hearing this speech overrides his hesitance to leave the comforts of the city, and he is eventually charmed by the "beautiful resting place" under the plane tree to which Phaedrus leads him, remarking that Phaedrus has "been the most marvelous guide" (230b-c). Phaedrus taunts Socrates nonetheless, commenting that he seems clearly out of his element; he says: "And you, my remarkable friend, appear to be totally out of place. Really, just as you say, you seem to need a guide, not

to be one of the locals. Not only do you never travel abroad—as far as I can tell, you never even set foot beyond the city walls" (230c-d). Socrates's seeming lack of familiarity with his environment and sense of being "totally out of place (*atopotatōs*) in the country" may be read as speaking to the emotions and attachments humanistic geographers see as necessary for acquiring a sense of place (Nehamas and Woodruff *ix*).

Along similar lines, Nedra Reynolds draws on this initial scene of Phaedrus and Socrates's stroll to illustrate how memory, embodiment, spatiality, and place are central to and inhabit "acts of composing" (2). Reynolds notes that "places evoke powerful human emotions," and thus Socrates's experience of place will likely shape his receptiveness to new ways of thinking (2). She argues that theories of writing must "reflect [a] deeper understanding of place" because sense of place is crucial to writing practices. Writing is achieved, Reynolds argues, not solely by the limitations or crossing of boundaries, but also through everyday practices and structures of feelings that have become engrained, habitual, and embodied (2). Feminist geographers similarly claim that "sense of place" forms in response to a combination of contexts and relations, "rather than being enacted by boundaries and through exclusions" that dictate who belongs and who doesn't (qtd. in McDowell and Sharp 201).

4. As Harley has argued, the field of cartography has traditionally been rooted in an objective, standardized set of practices that purport to convey accurate and correct models for ways of knowing. Even prior to the advent of the spatial sciences, it has been the goal of cartographic practice to "produce a 'correct' relational model of the terrain. [. . .] Similarly, the primary effect of the scientific rules was to create a 'standard'—a successful version of 'normal science'—that enabled cartographers to build a wall around the citadel of the 'true' map" (234–35). Contemporary understandings of cartography as a cultural practice are owed largely to recent debates within the field of GIS about the knowledge-making power of maps. In the mid-1990s, for example, the landmark essay collection, *Ground Truth,* noted that any definition of geographic information systems (GIS) or the digital maps produced by GIS should acknowledge the idea that the technology is "part of a contemporary network of knowledge, ideology, and practice that defines, inscribes, and represents environmental and social patterns within a broader economy of signification that calls forth new ways of thinking, acting, and writing" (Pickles, "Representations" 4). Nadine Schuurman chronicles three "waves" of critique in the history of GIS, of which the publication of *Ground Truth* played a large role. These three waves, she notes, reflect both a changing zeitgeist in the GIS landscape and shifting views "on the part of the critic" (Schuurman 570). The first two waves of GIS critique were often fraught with the suggestion that GIS was a positivist, value-neutral technology which functioned as "a mere *tool* of knowledge production" (Propen, "Critical GPS" 133). Later critiques understood GIS as socially constructed (see Harvey and

Chrisman, 1998; Harvey, 2001), as well as necessarily informed by a feminist theory approach (see Schuurman and Pratt, 2002; Kwan, 2002; and McLafferty, 2002).

5. Semiotics involves "the study of signs and signals, sign systems, and sign processes" (Moriarty 20). Signs typically stand for something else, such as "an object or concept" (21). There are two main approaches to the study of signs: that forwarded by American philosopher Charles Sanders Peirce, and that of Ferdinand de Saussure. Saussure's semiology takes a linguistic approach that focuses on "the two-part sign relationship between a signifier and its signified" (Moriarty 12). Saussure's theory is sometimes thought to be less useful than that of Peirce's, for as feminist geographer Gillian Rose writes, "Saussure had rather a static notion of how signs work and was uninterested in how meanings change and are changed in use. Other writers wonder whether a theory based on language can deal with the particularities of the visual" (77). Peirce, on the other hand, developed an arguably more useful "typology of signs" that includes his well-known icon, index, and symbol triad (Rose 78). Peirce's typology has been called "richer" than Saussure's because his understanding of signs "is differentiated by the way in which the relation between the signifier and signified is understood" (Rose 78). Briefly put, the icon is a sign that bears obvious similarity to its object (e.g. the map of California bears a direct likeness to the shape of the state of California); an index is a sign that is related to its object through a causal or sequential relationship (e.g. the printout from an electrocardiogram is a visual representation that results from the electrocardiogram's having been conducted); a symbol is a sign that correlates with its object by way of conventional or "natural" characteristics; convention may be socially or culturally dependent, and varies within different contexts (e.g. a Christmas tree is a symbol for Christmas) (Moriarty 21; see also Liszka, 1996). Certainly, as Marguerite Helmers and Charles A. Hill describe, "Peirce's distinctions are useful to rhetoricians because they establish a formal terminology for considering different types of imagistic sign systems, from representational, through diagrammatical, to allegorical" (15). Nonetheless, while Peirce's triad does acknowledge and rely on social and cultural contexts, a discussion based solely on the symbolic meaning of graphical features still does not fully allow for an understanding of the ways in which an artifact engages the body, or the ways in which maps or other spatially-based, material artifacts engage or contextualize the body within their discursive fields. This limitation may be due in part to the fact that, as Helmers and Hill also note, Peirce's "background in the natural sciences caused him to search for a logical, scientific method that would not be confused by what he termed 'beliefs.'" (15). Rose articulates a similar critique in *Visual Methodologies,* noting that "semiological studies focus on the image itself and there is thus little attention paid to audiencing and little concern for reflexivity" (99). Thus, while these semiotic concepts do provide a useful

terminology for analyzing different types of graphical features that function within sign systems such as the map, they function more appropriately as a point of entry into or as supplemental to analysis that then requires an interpretive lens more attuned to the nuanced understandings of how visual-material artifacts function within broader cultural contexts and reflect embodied cultural practices.

6. Of course, maps are not the only rhetorical, cultural artifacts through which selectivity functions as a component of meaning-making. As Prelli describes in his introduction to the collection *Rhetorics of Display:*

> Whether constituted through vocal enunciation, textual inscription, visual portrayal, material structure, enacted performance, or some combination, rhetorical study of displays proceeds from the central idea that whatever they make manifest or appear is the culmination of selective processes that constrain the range of possible meanings available to those who encounter them. (2)

In other words, Prelli says, "meanings manifested rhetorically through display are functions of particular, situated resolutions of the dynamic between revealing concealing. Put directly, whatever is revealed through display simultaneously conceals alternative possibilities; therein is display's rhetorical dimension" ("Rhetorics" 2). I might add here that, in addition to the rhetorical act of revealing and concealing inherent in the representation or display, the influence of the display on the bodies that inhabit or navigate it is also a key component of its rhetorical work.

7. It is on this issue of mapping as cultural practice that Crampton finds limitation in Harley's understanding of the map as an ideological text. In addition to his acceptance that "rhetoric is part of the way all texts work" (242), Harley notes that "[p]ower comes from the map and it traverses the way maps are made. Maps are a technology of power, and the key to this internal power is cartographic process" (244). Crampton feels this view is limited:

> While Harley [. . .] provided an examination of the political in mapping, that is to say, how maps are employed as political documents, this totally evades the question of how mapping *necessarily produces* the political [. . .]. On the Harleian agenda the political is assumed unproblematically and the history of mapping is privileged from *within* a discipline rather than questioning ("problematizing") the condition of possibility for thinking, a spatial politics. (52)

8. NASA officially credits Photo 22727 to the entire crew of Apollo 17: Eugene A. Cernan, Ronald E. Evans, and Harrison H. Schmitt (Hartwell).

9. Interestingly, as Cosgrove also notes, "in pre-Copernican cosmology the ascent of the soul through the spheres dissolved self into the harmony of a single creation precisely to escape the material rootedness of the dwelling body, fixed in place" (*Apollo's Eye* 263).

CHAPTER TWO

1. The pictorial turn, says Mitchell, is a response to *iconophobia,* or "the general anxiety of linguistic philosophy about visual representation" (12). It described the shift that arose in response to the earlier "need to defend 'our speech' against 'the visual'" (12–13). Finnegan describes the pictorial turn as rooted "in the stories of philosophy that take up visuality," citing "Peirceian semiotics, Goodman's philosophical aesthetics, Derrida's grammatology, and the Frankfurt School's critique of mass culture" as four such stories ("Review Essay" 236).

2. Moreover, in discussing the creation, distribution, and propagation of multimedia texts, Jonathan Alexander suggests that we must "take advantage of media convergence to re-examine, re-assess, and re-vision some of the most important convergences—and relationships—of our teaching and composing lives" (6). To do so, he says, "is necessary if we are to understand the many ways in which what passes as literate practice itself changes as relationships among media, producers of media, consumers of media, composers of multimedia, and composers and their composing tools all shift, collide, converge, and change" (6). The GPS then functions as a rhetorical artifact that exemplifies media convergence through its interactivity and the ability to program a route and register one's movements through a highly symbolic representation that has corporeal impact in the world.

3. We may note that Finnegan's notion of the image vernacular implicitly works against such a loss of agency by understanding "images as inventional resources for argument" ("Recognizing Lincoln" 63).

4. Foucault writes that heterotopias may be located in "reality" and that they are "formed in the very founding of society" ("Of Other Spaces" 24). We may read Foucault's notion of a "real" site, or the "reality" of place as consistent with his understanding of truth as anchored in the social world and never "outside power, or lacking in power" ("Truth and Power" 131). That is, Foucault writes:

> Truth is a thing of this world: it is produced only by virtue of multiple forms of constraint. And it induces regular effects of power. Each society has its régime of truth, its 'general politics' of truth: that is, the types of discourse which it accepts and makes function as true; the mechanisms and instances which enable one to distinguish true and false statements, the means by which each is sanctioned; the techniques and procedures accorded value in the acquisitions of truth; the status of those who are charged with saying what counts as true. (131)

This conceptualization of truth has implications for visual-material rhetoric and the idea that heterotopic sites may be located in reality. First, to understand societies as accepting particular discourses and making them "function as true" (131) is not incompatible with Blair's notion that "material phenomena" are "historically and contextually accreted understandings" that assume a "natural reality" (50). The reality in which visual-material artifacts are enmeshed, then, is discursively produced and socially contextualized. As Blair writes, "rhetoric occurs in a pedestrian world and exerts its most important consequences in the realm of human affairs" (51). Thus, she explains, "we must be mindful of the *social* world, which would include the only meaningful characterizations we have available of the 'natural world in which we all reside'" (51). To understand, for example, a map or a commemorative sculpture as a visual-material heterotopic artifact that functions in reality is to recognize not only its rhetorical power in the social, "pedestrian" world, but also the mechanisms that accorded that artifact its rhetorical status in a social reality in the first place.

5. For an additional discussion of the cemetery in particular as a heterotopia, see Wright, 2005.

6. Foucault's understanding of heterochronies as attending to the timelessness and timeliness of heterotopic sites also bears an interesting analogue to Halloran and Clark's discussion of the participation of the epideictic in the shaping of collective memory. The epideictic, they write, "is about something out of time [. . .] but it must happen in time. [. . .] [It] attempts to illuminate the timeless in some specific here and now. The collective memory that is nourished by epideictic aspires to the state of timelessness, but it subsists in time. It feeds on rituals, festivals, and observances that we attend at specified times, after which we return to everyday activities" (154). In other words, memory practices rely on the always shifting, contextualized, embodied experiences of particular temporalities.

7. Gregory Clark describes a similar idea when he expresses Kenneth Burke's acknowledgement of "the wordless rhetorical power that is wielded by the physical presence of things themselves" (28).

CHAPTER THREE

1. The mill owners were thought to have an underlying goal with the publication of the *Lowell Offering.* The women who worked in the mills formed what were called "improvement circles," which functioned as small clubs in which they wrote and shared creative work. These improvement circles were encouraged by the mill owners and local clergymen, largely because they directed the mill operatives' attention toward cultural activities rather than toward "complaining about their conditions in the mills and acting together to remedy them. [. . .] The major emphasis in these issues was to dispel the notion that factory work was degrading and that the mill operatives were exploited" (Foner 26). The mill operatives, however, did manage to subtly subvert the push to write and publish only idyllic portraits of mill

life. While some of the fiction and poetry published in the *Lowell Offering* indeed reveals the operatives' dissatisfaction with their working conditions, these pieces generally ended, however, "in the same escapism that character-ized the main bulk of the contributions" (26). Thus, contributions in the *Lowell Offering* often contained contradictory narratives and subsequently may be read as sending mixed messages regarding life in the mills. That is, narratives published in the *Lowell Offering* often shift from romanticizing life in the mills, to contemplating the tedious nature of the work, to rationalizing or justifying the lives of the Mill Girls.

2. To better inform my analysis of the Lowell Mills National Histori-cal Park, I first conducted a close reading of the narratives published by the Mill Girls in the *Lowell Offering.* In total, five volumes of the *Lowell Offering* were published between 1840 and 1845 (Farley). While these volumes are not extremely long, they are extraordinarily rich with material about the lives of the Mill Girls. I read each of these volumes, taking specific interest in the Mill Girls' descriptions of the mills and boardinghouses, any other related descriptions of their physical surroundings, and how these spaces affected their minds and bodies. Over the course of reading and analyzing these nar-ratives, I found that the following themes emerged: 1) descriptions of the mill as sacred space: expressions of wonder and awe mixed with fear and submis-sion; 2) attempts to romanticize or rationalize life in the mills; 3) descriptions of the mills as neat and clean; 4) plants and flowers in the mills (related to the theme of romanticizing and neatness); 5) the effect of the mills and board-inghouses on the body; and 6) the presence of loud noises in the mills. These themes each pertain to the impact of the mills on the bodies of the Mill Girls.

Narratives published in the *Lowell Offering* have also been anthologized by Benita Eisler in a collection titled *The Lowell Offering: Writings by New England Mill Women (1840–1845).* Eisler's is a secondary source that com-piles many narratives from the original publications. She organizes all of the Mills Girls' narratives thematically, into the following categories:

> Mill and Boardinghouse: The New Community
> Continuing Education: "A Glimpse of Something Grand Before Us"
> Looking Back: Nature, Family, and Childhood
> Choice and Conflict: The Cost of Independence
> Caste and Class: Mill Job or Marriage
> Changes: Reform, Regret, and a Vision of the Future

Eisler does not provide a formal rationale for why she chooses the categories she does, though her groupings are, overall, consistent with the themes upon which the Mill Girls generally focused their writings, as well as the chronol-ogy of the rise and fall of the mills in Lowell.

It would be inaccurate and naïve to state simply that the *Lowell Offering* provided the female workers at the Lowell Mills with a forum for creative expression in which they could also speak objectively about their working conditions. A close reading of the narratives published in the *Lowell Offering* reveals the pieces to be far more nuanced than this, functioning subversively to provide what are implicitly visual-material rhetorical accounts of the mills and the lives of the female operatives who labored there. Notable too is that the archives available at the University of Massachusetts Lowell Online Library describe many of the Mill Girls' narratives to be fictional accounts of mill life; however, this does not alter how I have understood the nature of these pieces. In fact, the Mill Girls' fictional accounts of mill life may have even had the effect of allowing these writers to distance themselves enough from their own experiences to be able to create stories that truly epitomize the aspects mill life which they felt were most important to convey to readers. For, as Tim O'Brien notes: "By telling stories, you objectify your own experience. You separate it from yourself. You pin down certain truths. You make up others" (158). While it makes sense that the Mill Girls' narratives are indeed informed by their daily experiences of life in the mills, it also makes sense that fictionalized accounts allow for some distance that better enables them to make choices that best reflect the nuances of life in the mills. Any quotations from narratives that I invoke in chapter three are then grounded in the idea that these fictional accounts are likely based on the Mill Girls' own personal experiences. It is therefore this combination of their own memories and firsthand experiences, coupled with their imagination and written narratives that work together to tell a complete story: "In a story, memory, imagination, and language all combine to create the illusion of aliveness" (O'Brien 230). Indeed, it is this combination of memory, imagination, and language that together form the Mill Girls' complex and engaging narratives.

3. I visited the park four times between 2003 and 2009, over several seasons and in different weather conditions. During these visits, I conducted observations of park visitors as they navigated green spaces and engaged with the various sculptures. I did not approach or initiate conversations with visitors so as not to influence or intrude on their experience. I recorded my observations in written field notes and also took photographs of park sculptures and green spaces, some of which are included and reproduced in this chapter. During my 2007 visit, I also interviewed a senior park ranger who has worked at the park for nearly twenty years. While interviewing was not my primary method of data collection in my analysis of the park, it did help to contextualize my observations and interpretations of the park and its artifacts.

4. The legislature ruled that it did not have "power to determine 'hours of work,'" and that the length of the workday had to be decided between the corporation and the workers ("Sarah George Bagley"). Pressure on the cor-

poration was great, however, and in 1847 they finally agreed to shorten the workday, but only by thirty minutes.

5. The *Lowell Offering* is a genre quite different from the *Voice of Industry*. The labor reform rhetoric in the *Voice* was much more charged and direct. It often included sections of proposed legislation, and addressed a readership beyond that of the antebellum Mill Girl, speaking also to the influx of immigrant workers in the late 1800s. Also pertinent are how the sculptures analyzed in this chapter may be interpreted as reflecting the tension, disorientation, and chaos of the mills during that time, and the effect of this chaos on both the Mill Girl and the park visitor. As chapter three describes, I believe that these sculptures very much reflect the mills' influence on the Mill Girls' bodies, even though they ostensibly commemorate a period just after the *Lowell Offering*. I feel these memorials can do both (commemorate the period before and after) because the labor reform movement builds upon and was borne out of the oppressive working conditions of the mills in the early to mid-1800s. So then, relevant in my mind are both the narratives written by the Mill Girls in the period preceding the labor reform movement, and the sculptures commemorating the female laborers of Lowell more generally, which I believe do gesture back to the oppressive mill conditions of the early 1800s.

6. Throughout the LMNHP, visitors will come across various interpretive plaques that describe the site in terms of its industrial heritage, specifically its relationship to the rise of hydroelectric power. In this chapter, however, I invoke and describe such plaques only as they pertain to or help contextualize the sculptures and art installations that I analyze. It is worth noting, however, that these plaques do supplement the green spaces and sculptures with textual information, and in doing so, serve as part of the park's intertext. Readers will also note that these plaques each contain a small image of the river and the Lowell canal system in the upper left-hand corner, similar to the larger image that comprises the wayfinding signs. The cartographic representation of the river and canal system in the interpretive plaques does not explicitly function in the service of wayfinding; rather, these smaller reproductions function more symbolically to link the interpretive plaques with the wayfinding signs, thus weaving them intertextually into the park's sign system.

7. As Blair says, "Even the bare materiality of a memorial site does not guarantee that it is the same text on a cloudy day as on a sunny one, on a crowded day as when almost deserted, at dawn as at midday. In fact, its capacity to be engaged physically actually determines its extreme mutability" (39). As a case in point, during my first trip to the LMNHP, the ground was covered with snow and so I did not notice the *Steps* sculpture. Upon returning to the park during a subsequent trip, I was initially surprised by its pres-

ence in the landscape before making the connection as to why I would have missed it prior.

8. A librarian at the Center for Lowell History at the University of Massachusetts, Lowell, thought this poem was written by Sarah Bagley; however, the park ranger speculated that the poem was written by Lucy Larcom. I am still unsure of the poem's actual name.

9. A librarian at the Center for Lowell History and the park ranger both confirmed this.

10. While I do not conduct an analysis of this museum within this chapter, it is worth noting that I did visit the museum in order to help contextualize my observations of the park. Before entering the room, a park ranger provides visitors with ear plugs to help reduce the noise level as they walk through the exhibit. The sound of the looms, even with ear plugs (which the Mill Girls did not wear) is deafening, and it is easy to imagine the hearing loss and tinnitus incurred by the mill operatives. The park ranger reminded me also that this exhibit only contains twenty functioning looms, whereas a typical room contained two hundred looms. Visitors to this exhibit therefore see firsthand, and very clearly, how this work environment impacted the bodies of the Mill Girls.

11. Here, the park ranger indeed identifies intent in the design and placement of the concert stage, though because I trust in her expertise, I feel comfortable relaying this acknowledgment of the artist's possible intent.

Chapter Four

1. See discussion of data collection and organization in this chapter, and note 9, below, for an explanation of the key with which I refer to quotes from interview participants in this chapter. I do not apply the additional descriptor "GPS user" in citations from interview participants throughout the rest of the chapter; I only do so in their initial usage because I have not yet described my method of data collection and so these codes would have little meaning for the reader without this additional context.

2. Due to scheduling difficulties, four interviews were conducted over email.

3. Delimiting the sample in this way not only allowed for easy access to a local population but also extended the sample beyond just that of full-time faculty, thereby preventing a more insular or limited sample of solely academic users. Additionally, it helped to set some parameters around what could potentially become a very broad population consisting of everyday GPS users. My decision to delimit the parameters of my sample to faculty, staff, and their family members at the college was based on the sample used by Barry Brown and Eric Laurier in their ethnographic study of the practices of city tourists in Glasgow and the implications of those practices for the design of

electronic mapping systems (1). To this end, Brown and Laurier, respectively from the Department of Geography and the Department of Computer Science at the University of Glasgow, "recruited groups of visitors to the city from friends and family" of their university's staff (2). Brown and Laurier presumably excluded university faculty and staff from their sample because, in their case, they wanted tourists to be wholly unfamiliar with the city, and faculty and staff who work at the university would already have local familiarity, whereas friends and families would presumably visit from other locales. Their choice to recruit participants with ties to a particular institution informed my rationale for understanding my participant pool as a starting point that could help delimit an otherwise potentially broad sample population of everyday GPS users. That is, Brown and Laurier's sample of Glasgow tourists could have been potentially quite broad without their having also defined "Glasgow tourists" as those people who were not only tourists to the area but also friends and family of their university's staff. Likewise, I defined "everyday GPS users" as those people who were not only GPS users in their own right but also either faculty and staff at the college, or family members of any faculty or staff at the college.

4. Readers will note that I use the plural pronoun "they" or "their" when referring to a GPS user. Since participant responses did not tend to fall along gendered lines, I did not want to cause any confusion or distraction by preceding a quotation with a gendered pronoun, potentially leaving readers to wonder whether any significance ought to be inferred through its use.

5. I also viewed the library's borrowing program as a way of providing those curious about the GPS with access to the technology without their having to invest financially in the device. In fact, several participants indicated that they'd been curious about what it would be like to use a GPS, but did not want to purchase one without first having the opportunity to try one out. Thus, they saw my study as a good excuse to "test drive" a GPS, and the library's program allowed them to do so. Because the library's borrowing program is open only to college faculty and staff, all participants in this study are employees, and none were students. The library was enthusiastic about helping out with this study and I am of course grateful for their assistance.

6. I provided interview questions in advance as part of the consent form. The interviews generally proceeded in accordance with my questionnaire (see Appendix A), with some departures given participants' responses, or when I chose to follow up on any markers.

7. While I could have used the interview questions as an organizing framework to begin the coding process, I chose instead to undertake a close reading of the interview transcripts, often reading them multiple times, to identify any recurrent topics or themes that emerged. In a separate document, I then created a coding outline, which contained a list of these themes, or categories and sub-categories that emerged. I then copied quotations from

each transcript into the pertinent sections of the coding document. (See Appendix B for a table that contains my list of categories and subcategories.) Some categories and sub-categories in the coding document turned out to be more populated with quotations than others, and some categories did not turn out to be relevant to the direction taken in the final analysis, and are thus not represented in chapter four. Of course, as I coded, I was also thinking about Blair's theory of material rhetoric and Foucault's theory of heterotopias, and I would be remiss not to note that these ideas did inform my interpretation of the interview data to some extent; in other words, theory functioned as an interpretive tool in this sense. As I also note in chapter four, the major themes that emerged during the coding of the interviews generally drove my organization of the chapter—this is because I wanted to allow the experiences and voices of the participants to truly be reflected in my analysis. The result is then an organizational framework for the chapter that not only reflects the themes and trajectories of participants' responses but also resonates with an understanding of GPS use as visually and materially rhetorical and geographically sensitive. In this way, I take my approach here to be aligned with a grounded theory methodology, in which "concepts are formulated and analytically developed," and always reflective of "multiple perspectives" (Strauss and Corbin 173). That is to say, the qualitative researcher will always carry with them both personal knowledge or experience, as well as the knowledge or voices of participants or "actors studied"; thus, the "interplay between researcher and the actors studied [. . .] is likely to result in some degree of reciprocal shaping" (173). This idea of reciprocal shaping, again, informed not only my coding of the interview transcripts but also the organizational framework of the chapter.

8. Nonetheless, to further contextualize my research, I did familiarize myself with some of the studies related to usability and safety of in-car navigational devices and an ethnographic study of GPS use geared toward improving GPS design features. Much of this research is published within the fields of information science and human-computer interaction (HCI), and while the goals or theoretical bent of these studies often differ from the direction of this work, I invoke some of them when they provide insights or information pertinent to comments made by participants in this study (see Blanco et al., 2006; Burnett et al., 2004; Leshed et al., 2008; and Noel et al., 2005).

9. In this chapter, I identify the speaker of a quote according to the following key, which allowed me to track responses on a macro-level yet maintain participant confidentiality: If the participant borrowed a GPS from the library, I refer to them with a "B" (in one case, a participant borrowed a GPS from a family member); if the participant owns a GPS, I refer to them with an "O"; if they are college staff or administration, I include an "SA"; if they are adjunct faculty, I include an "A"; if they are a full-time faculty member, I

include an "FT"; and if they are a family member of college faculty or staff, I include an "FM." Numbers are assigned randomly.

10. While GPS devices have countless applications and myriad uses, this chapter focuses on its use for purposes of in-car navigation. When used as such, the GPS, while displaying information in a virtual environment, is not necessarily functioning in the capacity of, say, a video game. Nonetheless, it bears noting that some users did note a potential interest in using the GPS for the purposes of activities such as geocaching or other forms of play. Thus the door remains open for research related to GPS "play" and its implications for the study of visual-material rhetorics.

11. I should note here that the consent form signed by participants stated that participation in the study was completely voluntary. If at any time during the study participants became uncomfortable or did not wish to continue using the GPS, they were of course free to terminate their participation without any repercussions.

12. While recent scholarship in visual rhetoric and composition studies has been attentive to forms of visual, material, and digital rhetoric, it bears noting that scholars have also argued for a better integration of sound in the study of multimodal composition (Shipka 371).

13. According to Communication Studies scholars Kristine L. Nowak and Christian Rauh, "social cognition theory argues that the ability to identify anthropomorphic characteristics and categorize objects in the environment as humans, animals, or objects is a basic human cognitive function." Further, they note that "[o]bjects, animals, and humans form the basic social categories to which people assign the things they encounter." They understand anthropomorphism as "the attribution of human form or other human characteristics to any nonhuman object." In their study of the avatar, Nowak and Rauh "examine anthropomorphism only in terms of human morphology, or appearance, and not behavior," and thus they focus more so on the visible image of the avatar and its influence on online perceptions of anthropomorphism ("Influence of the Avatar"). In this chapter, anthropomorphism comes into play most visibly in participants' attribution of human characteristics to the non-human GPS; that is, participants often described their responses to the GPS's voice cues as though they were engaging in a dialogue with another human. Conversely, participants would sometimes describe the GPS as talking to them, particularly when describing its "recalculating" cues.

CHAPTER FIVE

1. Since 2000, mounting evidence has shown that low-frequency active sonar (LFA), used to detect submarines in matters of national security, can be deadly to marine mammals. LFA systems can "operate at more than 235 decibels, producing sound waves that can travel across tens or even hundreds

of miles of ocean." According to the National Resources Defense Council, "[d]uring testing off the California coast, noise from the Navy's main low-frequency sonar system was detected across the breadth of the northern Pacific Ocean." The Navy's LFA sonar "can retain an intensity of 140 decibels— a hundred times more intense than the level known to alter the behavior of large whales." In 2000, four different whale species were found stranded along beaches in the Bahamas. According to the NRDC, "[a]lthough the Navy initially denied responsibility, the government's investigation established that mid-frequency sonar caused the strandings." The sonar results in physical trauma to the whales, "including bleeding around the brain, ears and other tissues and large bubbles in their organs." As the NRDC describes, "These symptoms are akin to a severe case of 'the bends'—the illness that can kill scuba divers who surface quickly from deep water. Scientists believe that the mid-frequency sonar blasts may drive certain whales to change their dive patterns in ways their bodies cannot handle, causing debilitating and even fatal injuries" (NRDC "Lethal Sounds").

2. For a more detailed description of the Longhurst biogeographical provinces, see chapter six of Alan R. Longhurst's *Ecological Geography of the Sea* (2006). The Longhurst biomes are also used in research related to fisheries management (see Pauly et al., 2000).

3. The PowerPoint presentation was created for the public by NOAA's National Marine Fisheries Service in 2002, during the time that the NMFS issued its Final Rule on the Navy's use of LFA sonar. Readers may view the NMFS's map, which is found on page 7 of the NMFS presentation called "MMPA Small Take Authorization," by visiting the following URL on the NOAA Fisheries Service website: <http://www.nmfs.noaa.gov/pr/pdfs/permits/mmpa_small_take.pdf> Readers may also learn more about the history of LFA sonar by visiting the following URL on the NOAA Fisheries Service website: <http://www.nmfs.noaa.gov/pr/acoustics/surtass.htm#pubs>.

4. Though it was this GIS specialist who originally created the map in Figure 46, the map belongs to the NRDC, and so I refer to it as such in this chapter. Between 2003 and 2005, I kept in contact with the GIS manager regarding questions and clarifications about the case. I also received permission from the NRDC to reprint their map, and since it was used in open court, it is considered open to the public.

5. Because the map included in the NMFS's PowerPoint presentation is not introduced or mentioned in the Opinion of the Court granting the preliminary injunction or the Summary Judgment, I do not treat it as though it were directly or explicitly involved in the court case, as I have no evidence that it was. Rather, in the Opinion of the Court and in the Summary Judgment, I focus on the treatment of the NRDC's map as "Exhibit A" in the preliminary injunction hearing, and the discussion of the NMFS's use of the Longhurst biomes in forwarding their claims about LFA use and specified

geographical region. Because the NRDC's map was based on the map in the NMFS's public PowerPoint presentation, however, I am interested in the relationship between these two artifacts and the implications of this relationship as it pertains to the study of visual-material rhetorics.

6. It is worth noting that the surrounding presentation includes information about the areas of ocean where LFA sonar will be prohibited. These areas are referred to as "Offshore Biologically Important Areas," and include areas "within 12 nmi [nautical miles] of any coast or island" (Hollingshead 11). The presentation also notes that LFA sonar will be prohibited in Arctic waters, so as not to adversely impact subsistence hunting in those regions (13). The map on page 7 of the presentation does not explicitly depict areas of ocean deemed Offshore Biologically Important Areas, and does not appear to include Arctic waters (with the possible exception of biogeographic area 50, which the NRDC's map does not portray as affected by LFA sonar), and so the map appears to be consistent with subsequent statements about where the sonar will be prohibited. While these points provide some context for this analysis, they do not alter a visual-material rhetorical analysis of the actual "MMPA Small Take Authorization" map.

7. Here, Haraway refers to Latour's "Great Divides" as helping to distinguish "between what counts as nature and as society, as human and as nonhuman" (9).

Works Cited

Alexander, Jonathan. "Media Convergence: Creating Content, Questioning Relationships." *Computers and Composition* 25.1 (2008): 1–8. *ScienceDirect*. Web. Aug. 2008.

Anderson, Paul B. "Behrmann Projection." *Map Projections*. Center for Spatially Integrated Social Science. Nov. 2002. Web. Dec. 2009.

—. "Mercator Projection." *Map Projections*. Center for Spatially Integrated Social Science. Nov. 2002. Web. Dec. 2009.

Authorized Deployment of LFA: 2002–2003. Map. Washington, D.C.: NRDC, 2002. Print.

Barton, Ellen. "Discourse Methods and Critical Practice in Professional Communication: The Front-Stage and Back-Stage Discourse of Prognosis in Medicine." *Journal of Business and Technical Communication* 18.1 (2004): 67–111. Print.

Barton, Ben F., and Marthalee S. Barton. "Ideology and the Map: Toward a Postmodern Design Practice." *Professional Communication: The Social Perspective*. Ed. Nancy Roundy Blyler and Charlotte Thralls. Thousand Oaks, CA: Sage, 1993. 49–78. Print.

Bartholomae, David, and Anthony Petrosky, eds. *Ways of Reading: An Anthology for Writers*. 6th ed. Boston: Bedford/St. Martin's, 2002. Print.

Berger, John. *Ways of Seeing*. London: BBC and Penguin, 1972.

Biesecker, Barbara. "Michel Foucault and the Question of Rhetoric." *Philosophy and Rhetoric* 25 (1992): 351–364. Print.

—. "Remembering World War II: The Rhetoric and Politics of National Commemoration at the Turn of the 21st Century." *Visual Rhetoric: A Reader in Communication and American Culture*. Ed. Lester C. Olson, Cara A. Finnegan, and Diane S. Hope. Thousand Oaks, CA: Sage, 2008. 157–174. Print.

Bizzell, Patricia, and Bruce Herzberg. *The Rhetorical Tradition: Readings from Classical Times to the Present*. Boston: Bedford/St. Martin's, 1990. Print.

Blair, Carole. "Contemporary U.S. Memorial Sites as Exemplars of Rhetoric's Materiality." *Rhetorical Bodies*. Ed. Jack Selzer and Sharon Crowley. Madison: U of Wisconsin P, 1999. 16–57. Print.

Blair, Carole, Marsha S. Jeppeson, and Enrico Pucci, Jr. "Public Memorializing in Postmodernity: The Vietnam Veterans Memorial as Prototype." *Quarterly Journal of Speech 77* (1991): 263–288. Print.

Blair, Carole, and Neil Michel. "Commemorating in the Theme Park Zone: Reading the Astronauts Memorial." *At the Intersection: Cultural Studies and Rhetorical Studies.* Ed. Thomas Rostek. New York: Guilford Press, 1999. 29–83. Print.

—. "Reproducing Civil Rights Tactics: The Rhetorical Performances of the Civil Rights Memorial." *Visual Rhetoric: A Reader in Communication and American Culture.* Ed. Lester C. Olson, Cara A. Finnegan, and Diane S. Hope. Thousand Oaks, CA: Sage, 2008. 139–156. Print.

Blakesley, David. "Defining Film Rhetoric: The Case of Hitchcock's *Vertigo.*" *Defining Visual Rhetorics.* Ed. Charles A. Hill and Marguerite Helmers. Mahwah, NJ: Lawrence Erlbaum, 2004. 111–134. Print.

—. "What is Visual Rhetoric, and What Is Its Tradition? Position Statement: Alliance of Rhetoric Societies." Purdue University. Sept. 2003. Web. Nov. 2009.

Blanco, Myra, Wayne J. Biever, John P. Gallagher, and Thomas A. Dingus. "The Impact of Secondary Task Cognitive Processing Demand on Driving Performance." *Accident Analysis and Prevention* 38 (2006): 895–906. *Academic Search Premiere.* Web. June 2009.

Brooke, Collin Gifford. "Forgetting to be (Post)Human: Media and Memory in a Kairotic Age." *JAC* 20.4 (2000): 775–795. Print.

Brouwer, Dan. "The Precarious Visibility Politics of Self-Stigmatization: The Case of HIV/AIDS Tattoos." *Visual Rhetoric: A Reader in Communication and American Culture.* Ed. Lester C. Olson, Cara A. Finnegan, and Diane S. Hope. Thousand Oaks, CA: Sage, 2008. 205–226. Print.

Brown, Barry, Eric Laurier, and Hayden Lorimer. "Driving and 'Passengering': Notes on the Ordinary Organization of Car Travel." *Mobilities* 3.1 (2008): 1–23. *Google Scholar.* Web. Aug. 2008.

Burnett, G. E., S.J. Summerskill, and J.M. Porter. "On-the-Move Destination Entry for Vehicle Navigation Systems: Unsafe By Any Means?" *Behaviour and Information Technology* 23.4 (2004): 265–272. *Illiad.* Web. June 2009.

Calvert, Kori, and Eugene H. Buck. "Active Sonar and Marine Mammals: Chronology with References." *CRS Report for Congress.* Congressional Research Service. Library of Congress. June 2005. Web. Jan. 2010.

Campbell, Karlyn Kohrs. "Agency: Promiscuous and Protean." *Communication and Critical/Cultural Studies* 2.1 (2005): 1–19. *Academic Search Premiere.* Web. July 2009.

Carr, Nicholas. "Is Google Making Us Stupid?" *The Atlantic.* July/Aug. 2008. Web. Sept. 2008.

Clark, Gregory. *Rhetorical Landscapes in America: Variations on a Theme from Kenneth Burke*. Columbia: U of South Carolina P, 2004. Print.

Cooper, Desiree, and Millie Jefferson. "Don't Ask the GPS: Interview with Mark Monmonier." *Weekend America: National Public Radio*. Mar. 2008. Web. Nov. 2009.

Corbin, Juliet M., and Anselm Strauss. "Grounded Theory Research: Procedures, Canons, and Evaluative Criteria." *Qualitative Sociology* 13.1 (1990): 3–21. Print.

Cosgrove, Denis. *Apollo's Eye: A Cartographic Genealogy of the Earth in the Western Imagination*. Baltimore: Johns Hopkins UP, 2001.

—. "Mapping Meaning." Introduction. *Mappings*. Ed. Denis Cosgrove. London: Reaktion, 1999. 1–23. Print.

—. "New World Orders." *New Words, New Worlds: Reconceptualising Social and Cultural Geography*. Ed. Chris Philo. Edinburgh: IBG Social and Cultural Geography Study Group, 1991. 125-130.

Cosgrove, Denis, and L. L. Martins. "Millennial Geographics." *Annals of the Association of American Geographers* 90 (2000): 97–103.

Crampton, Jeremy W. *The Political Mapping of Cyberspace*. Chicago: U of Chicago P, 2003. Print.

Crampton, Jeremy W., and John Krygier. "An Introduction to Critical Cartography." *ACME: An International E-Journal for Critical Geographies* 4.1 (2006): 11–33. Web. June 2009.

Crowley, Sharon. "Afterword: The Material of Rhetoric." *Rhetorical Bodies*. Ed. Jack Selzer and Sharon Crowley. Madison: U of Wisconsin P, 1999. 357–366. Print.

Cumming, Robert. *The Lowell Sculptures: One, Two, and Three*. 1990. Public Art Installation. Lowell Mills Natl. Historical Park, Lowell, MA.

Dana, P. H. "Global Positioning System Overview." *The Geographer's Craft Project*. 2000. Web. Sept. 2008.

De Certeau, Michel. *The Practice of Everyday Life*. Trans. Steven Rendall. Berkeley: U of California P, 1988.

Dickinson, Greg, and Casey Malone Maugh. "Placing Visual Rhetoric: Finding Material Comfort in Wild Oats Market." *Defining Visual Rhetorics*. Ed. Charles A. Hill and Marguerite Helmers. Mahwah, NJ: Lawrence Erlbaum, 2004. 259–276. Print.

Dickson, Barbara. "Reading Maternity Materially." *Rhetorical Bodies*. Ed. Jack Selzer and Sharon Crowley. Madison: U of Wisconsin P, 1999. 297–313. Print.

Dorling, Daniel, and David Fairbairn. *Mapping: Ways of Representing the World*. New York: Prentice Hall, 1997.

Dublin, Thomas. "Women, Work, and Protest in the Early Lowell Mills: 'The Oppressing Hand of Avarice Would Enslave Us.'" *Labor History* 16 (1975): 99–116. Print.

"Earth Day History." *History.com.* A&E Television Networks. 1996–2001. Web. August 2010.

Edwards, Janis L., and Carol K. Winkler. "Representative Form and the Visual Ideograph: The Iwo Jima Image in Editorial Cartoons." *Visual Rhetoric: A Reader in Communication and American Culture.* Ed. Lester C. Olson, Cara A. Finnegan, and Diane S. Hope. Thousand Oaks, CA: Sage, 2008. 119–138. Print.

Eisler, Benita. *The Lowell Offering: Writings by New England Women (1840–1845).* New York: W.W. Norton and Company, 1998. Print.

Faber, Brenton. "Toward a Rhetoric of Change: Reconstructing Image and Narrative in Distressed Organizations." *Journal of Business and Technical Communication* 12 (2008): 217–237. Print.

Farley, Harriet, ed. *The Lowell Offering. Vols. I–V.* Lowell, MA: Powers and Bagley, 1844. Print.

Fausch, Deborah. "The Knowledge of the Body and the Presence of History—Toward a Feminist Architecture." *Architecture and Feminism: Yale Publications on Architecture.* Ed. Debra Coleman, Elizabeth Danze, and Carol Henderson. New York: Princeton Architectural Press, 1996. 38–59. Print.

Finnegan, Cara A. "Doing Rhetorical History of the Visual: The Photograph and the Archive." *Defining Visual Rhetorics.* Ed. Charles A. Hill and Marguerite Helmers. Mahwah, NJ: Lawrence Erlbaum, 2004. 195–214. Print.

—. "Recognizing Lincoln: Image Vernaculars in Nineteenth-Century Visual Culture." Ed. Olson, Lester C., Cara A. Finnegan, and Diane S. Hope. *Visual Rhetoric: A Reader in Communication and American Culture.* Thousand Oaks, CA: Sage, 2008. 61–78. Print.

—. "Review Essay: Visual Studies and Visual Rhetoric." *Quarterly Journal of Speech* 90 (2004): 234–256. Print.

Flynn, Regina Robbins. "The World Beyond MapQuest." *The Chronicle of Higher Education.* 20 July 2009. Web. Jan. 2010.

Foner, Philip S., ed. *The Factory Girls: A Collection of Writings on Life and Struggles in the New England Factories of the 1840s by the Factory Girls Themselves, and the Story, in Their Own Words, of the First Trade Unions of Women Workers in the United States.* Chicago: U of Illinois P, 1977. Print.

Foucault, Michel. *The Archaeology of Knowledge.* Trans. A.M. Sheridan Smith. New York: Pantheon Books, 1972. Print.

—. *Discipline and Punish: The Birth of the Prison.* Trans. Alan Sheridan. NewYork: Vintage Books, 1977. Print.

—. "Of Other Spaces." Trans. Jay Miskowiec. *Diacritics* 16 (1986): 22–27. Print.

—. "Truth and Power." *Power/Knowledge: Selected Interviews and Other Writings 1972–1977.* Ed. Colin Gordon. Trans. Colin Gordon, Leo Marshall,

John Mepham, and Kate Soper. Brighton: Harvester Press, 1980. 109–133. Print.

Grabill, Jeff T. "Shaping Local HIV/AIDS Services Policy through Activist Research: The Problem of Client Involvement." *Technical Communication Quarterly* 9 (2000): 29–50. Print.

Gross, Alan G. "Toward a Theory of Verbal-Visual Interaction: The Example of Lavoisier." *Rhetoric Society Quarterly* 39.2 (2009): 147–169. Print.

Gurak, Laura J., and Amy Propen. Rev. of *Eloquent Images: Word and Image in the Age of New Media,* Ed. Mary E. Hocks and Michelle R. Kendrick. *Technical Communication Quarterly* 14 (2005): 101–109. Print.

Haas, Christina. "Materializing Public and Private: The Spatialization of Conceptual Categories in Discourses of Abortion." *Rhetorical Bodies.* Ed. Jack Selzer and Sharon Crowley. Madison: U of Wisconsin P, 1999. 218–238. Print.

Halloran, Michael S., and Gregory Clark. "National Park Landscapes and the Rhetorical Display of Civic Religion." *Rhetorics of Display.* Ed. Lawrence J. Prelli. Columbia: U of South Carolina P, 2006. 141–156. Print.

Handa, Carolyn. "Introduction to Part Five." *Visual Rhetoric in a Digital World: A Critical Sourcebook.* Ed. Carolyn Handa. Boston: Bedford/St. Martin's, 2004. 377–380. Print.

Haraway, Donna J. *When Species Meet.* Minneapolis: U of Minnesota P, 2008. Print.

Hariman, Robert, and John Louis Lucaites. *No Caption Needed: Iconic Photographs, Public Culture, and Liberal Democracy.* Chicago: U of Chicago P, 2007. Print.

Harley, J. B. "Deconstructing the Map." *Writing Worlds: Discourse, Text and Metaphor in the Representation of Landscape.* Ed. Trevor J. Barnes and James S. Duncan. New York: Routledge, 1992. 231–47. Print.

Harris, Leila, and Helen Hazen. "Rethinking Maps From a More-Than-Human Perspective: Nature-Society, Mapping and Conservation Territories." *Rethinking Maps: New Frontiers in Cartographic Theory.* Ed Martin Dodge, Rob Kitchin, and Chris Perkins. New York: Routledge Studies in Human Geography, 2009. 50–67. Print.

Hartwell, Eric. "Apollo 17: The Blue Marble." *Eric Hartwell's InfoDabble.* Apr. 2007. Web. Aug. 2010.

Harvey, Francis. "Constructing GIS: Actor Networks of Collaboration." *URISA Journal* 13 (2001): 29–38. Print.

Harvey, Francis, and Nick Chrisman. "Boundary Objects and the Social Construction of GIS Technology." *Environment and Planning A* 30 (1998): 1683–1694. Print.

Hayles, Katherine N. "Deeper Into the Machine: Learning to Speak Digital." *Computers and Composition* 19 (2002): 371–386. *Academic Search Premiere.* Web. Aug. 2008.

—. *How We Became Posthuman: Virtual Bodies in Cybernetics, Literature, and Informatics.* Chicago: U of Chicago P, 1999. Print.

—. *My Mother Was a Computer: Digital Subjects and Literary Texts.* Chicago: U of Chicago P, 2005. Print.

Helmers, Marguerite. *The Elements of Visual Analysis.* New York: Longman, 2005. Print.

—. "Framing the Fine Arts Through Rhetoric." *Defining Visual Rhetorics.* Ed. Charles A. Hill and Marguerite Helmers. Mahwah, NJ: Lawrence Erlbaum, 2004. 63–86. Print.

Helmers, Marguerite, and Charles A. Hill. Introduction. *Defining Visual Rhetorics.* Ed. Charles A. Hill and Marguerite Helmers. Mahwah, NJ: Lawrence Erlbaum, 2004. 1–24. Print.

Henderson, Kathryn. "The Visual Culture of Engineers." *The Cultures of Computing.* Ed. Susan Leigh Star. Oxford: Blackwell Publishers, 1995. 196–217. Print.

Hocks, Mary, and Michelle Kendrick, eds. *Eloquent Images: Word and Image in the Age of New Media.* Cambridge: MIT Press, 2003. Print.

Hollingshead, Ken. "MMPA Small Take Authorization Presentation." *NOAA Fisheries Service.* NOAA, 2002. Web. Nov. 2009.

Hope, Diane S. "Gendered Environments: Gender and the Natural World in the Rhetoric of Advertising." *Defining Visual Rhetorics.* Ed. Charles A. Hill and Marguerite Helmers. Mahwah: Lawrence Erlbaum Associates, 2004. 155–178. Print.

"In the Shadow of the Mills." Lowell Mills National Historical Park. Lowell, MA. Printed Sign.

Johnson, Jim [Bruno Latour]. "Mixing Humans and Nonhumans Together: The Sociology of a Door-Closer." *Ecologies of Knowledge: Work and Politics in Science and Technology.* Ed. Susan Leigh Star. Albany: SUNY P, 1995. 257–277. Print.

Josephson, Hannah. *The Golden Threads: New England's Mill Girls and Magnates.* New York: Duell, Sloan and Pearce, 1949. Print.

Kaufman, Mico. *Homage to Women.* 1984. Sculpture. Lowell Mills Natl. Historical Park, Lowell, MA.

Kennedy, Krista. "Textual Machinery: Authorial Agency and Bot-Written Texts in Wikipedia." *The Responsibilities of Rhetoric.* Ed. Michelle Smith and Barbara Warnick. Long Grove, IL: Waveland Press, 2010. 303-309. Print.

Kenney, Keith. "Building Visual Communication Theory by Borrowing from Rhetoric." *Journal of Visual Literacy* 22 (2002): 53–80. Print.

Kim, Loel, Amanda J. Young, Robert A. Neimeyer, Justin N. Baker, and Raymond C. Barfield. "Keeping Users at the Center: Developing a Multimedia Interface for Informed Consent." *Technical Communication Quarterly* 17 (2008): 335–357. *Academic Search Premiere.* Web. Aug. 2008.

Kimball, Miles A. "Cars, Culture, and Tactical Technical Communication." *Technical Communication Quarterly* 15 (2006): 67–86. *Academic Search Premiere.* Web. Aug. 2008.

Knievel, Michael. "Technology Artifacts, Instrumentalism, and the Humanist Manifestos: Toward an Integrated Humanistic Profile for Technical Communication." *Journal of Business and Technical Communication* 20 (2006): 65–86. *Academic Search Premiere.* Web. Aug. 2008.

Kostelnick, Charles. "Melting-Pot Ideology, Modernist Aesthetics, and the Emergence of Graphical Conventions: The Statistical Atlases of the United States, 1874–1925." *Defining Visual Rhetorics.* Ed. Charles A. Hill and Marguerite Helmers. Mahwah, NJ: Lawrence Erlbaum, 2004. 215–242. Print.

Kostelnick, Charles, and Michael Hassett. *Shaping Information: The Rhetoric of Visual Conventions.* Carbondale: Southern Illinois UP, 2003. Print.

Kovarik, Bill. "Environmental History Timeline." Radford University. N.d. Web. Aug. 2010.

Kress, Gunther, and Theo van Leeuwen. *Reading Images: The Grammar of Visual Design.* New York: Routledge, 1996. Print.

Kwan, Mei-Po. "Is GIS for Women? Reflections on the Critical Discourse in the 1990s." *Gender, Place and Culture* 9 (2002): 271–279. Print.

Lake, Randall, and Barbara Pickering. "Argumentation, the Visual, and the Possibility of Refutation: An Exploration." *Argumentation* 12 (1998): 79–93. Print.

Latour, Bruno. "Drawing Things Together." *Representation in Scientific Practice.* Ed. Michael Lynch and Steve Woolgar. Cambridge: MIT Press, 1990. 19–68. Print.

Leshed, Gilly, Theresa Velden, Oya Rieger, Blazej Kot, and Phoebe Sengers. "In-Car GPS Navigation: Engagement with and Disengagement from the Environment." *Proceeding of the Twenty-sixth Annual SIGCHI Conference on Human Factors in Computing Systems.* New York: Association for Computing Machinery, 2008. *Illiad.* Web. June 2009.

Lindeman, Neil. "Creating Knowledge for Advocacy: The Discourse of Research at a Conservation Organization." *Technical Communication Quarterly* 16 (2007): 431–451. Print.

Liptak, Adam. "Supreme Court Rules for Navy in Sonar Case." *The New York Times.* 12 Nov. 2008. Web. 5 Aug. 2009.

Liszka, James Jakob. *A General Introduction to the Semiotic of Charles Sanders Peirce.* Bloomington: Indiana UP, 1996. Print.

Longhurst, Alan. *Ecological Geography of the Sea.* New York: Academic Press, 2006. Print.

Lowell National Historical Park: 1978- 2008: 30 Years of Preservation and Innovation for Future Generations. Online Booklet. Lowell, MA: Harpers Ferry Center, National Park Service. June 2008. Web. Jan. 2010.

Lowell National Historical Park. Map. Lowell, MA: Harpers Ferry Center, National Park Service. July 2000. Web. Jan. 2010.

Lowell Public Art Collection. "Homage to Women." Lowell Mills Natl. Historical Park, Lowell, MA. Printed Sign.

—. "The Lowell Sculptures: One, Two, and Three." Lowell Mills Natl. Historical Park. Lowell, MA. Printed Sign.

—. "Industry Not Servitude." Lowell Mills Natl. Historical Park, Lowell, MA. Printed Sign.

—. "In the Shadow of the Mills." Lowell Mills Natl. Historical Park, Lowell, MA. Printed Sign.

Luijten, Joep. "Re: Greetings and an Inquiry about 'Battle Maps.'" Message to the author. 23 June 2003. E-mail.

—. "Re: Referencing the LFA Maps." Message to the author. 6 Jan. 2004. E-mail.

Marback, Richard. "Unclenching the Fist: Embodying Rhetoric and Giving Objects Their Due." *Rhetoric Society Quarterly* 38 (2008): 46–65. Print.

Massey, Doreen. *Space, Place, and Gender.* Minneapolis: U of Minnesota P, 1994. Print.

McDowell, Linda, and Joanne P. Sharp. *A Feminist Glossary of Human Geography.* New York: Oxford UP, 1999. Print.

McLafferty, Sara L. "Mapping Women's Worlds: Knowledge, Power and the Bounds of GIS." *Gender, Place and Culture* 9 (2002): 263–269. Print.

Miller, Carolyn R. "What Can Automation Tell Us About Agency?" *Rhetoric Society Quarterly* 37.2 (2007): 137–157. *Academic Search Premiere.* Web. July 2009.

Mitchell, W. J. T. *Picture Theory: Essays on Verbal and Visual Representation.* Chicago: U of Chicago P, 1994. Print.

Moberly, Kevin. "Composition, Computer Games, and the Absence of Writing." *Computers and Composition* 25 (2008): 284–299. *Academic Search Premiere.* Web. August 2008.

Monmonier, Mark. *How to Lie with Maps.* Chicago: U of Chicago P, 1996. Print.

—. *Mapping it Out.* Chicago: U of Chicago P, 1993. Print.

—. *Spying with Maps: Surveillance Technologies and the Future of Privacy.* Chicago: U of Chicago P, 2002. Print.

Montrie, Chad. *Making a Living: Work and Environment in the United States.* Chapel Hill: U of North Carolina P, 2008. Print.

Moriarty, Sandra. "The Symbiotics of Semiotics and Visual Communication." *Journal of Visual Literacy* 22.1 (2002): 19–28. Print.

Muckelbauer, John and Debra Hawhee. "Posthuman Rhetorics: 'It's the Future, Pikul.'" *JAC* 20.4 (2000): 767–774. *Academic Search Premiere.* Web. June 2009.

Mullen, Mark. "Collapsing Floors and Disappearing Walls: Teaching Visual and Cultural Intertexts in Electronic Games." *Writing the Visual: A Practical Guide for Teachers of Composition and Communication.* Ed. Carol David and Anne R. Richards. West Lafayette: Parlor Press, 2008. 221–242. Print.

Natl. Aeronautics and Space Administration. "View of the Earth Seen by the Apollo 17 Crew Traveling toward the Moon (AS17–148–22727)." 1972. *Natl. Space Science Data Center Photo Gallery.* NASA. Web. July 2010.

Natl. Marine Fisheries Service: Office of Protected Resources. *Sonar.* Web. Jan. 2010.

Natural Resources Defense Council (NRDC). "Lethal Sounds: The Use of Military Sonar Poses a Deadly Threat to Whales and Other Marine Mammals." *Natural Resources Defense Council: The Earth's Best Defense.* NRDC. 6 Oct. 2008. Web. Nov. 2009.

Natural Resources Defense Council (NRDC) v. Evans. US District Court for Northern California. Complaint for Declaratory and Injunctive Relief for Violation of Marine Mammal Protection Act, National Environmental Policy Act, Endangered Species Act, and Administrative Procedure Act. 2002. *FindLaw.com.* FindLaw, 2011. Web. Jan. 2010.

—. No. C-02–3805-EDL. US District Court for Northern California. Opinion and Order Granting Plaintiffs' Motion for a Preliminary Injunction. 2002 (a). Print.

—. No. C-02–3805-EDL. US District Court for Northern California. Opinion on Order and Cross-Motions for Summary Judgment. 2003. Print.

Negin, Elliott. "Battle Maps: In the War of Words, Sometimes a Picture Can Win the Day." *Onearth* (2003): 46. Print.

Nehamas, Alexander and Paul Woodruff. "Introduction." *Phaedrus.* Trans. Alexander Nehamas and Paul Woodruff. Cambridge: Hackett, 1995. ix–xlvii. Print.

Nelson, Gaylord. "Earth Day '70: What It Meant." *United States Environmental Protection Agency.* EPA, Apr. 1980. Web. Aug. 2010.

Nicoletti, L. J. "Mediated Memory: The Language of Memorial Spaces." *Writing the Visual: A Practical Guide for Teachers of Composition and Communication.* Ed. Carol David and Anne R. Richards. West Lafayette: Parlor Press, 2008. 51–69. Print.

Noel, Elliott, Blair Nonnecke, and Lana Trick. "Evaluating First-Time and Infrequent Use of In-Car Navigation Devices." *Proceedings of the Third International Driving Symposium on Human Factors in Driver Assessment, Training and Vehicle Design.* 2005. Web. June 2009.

Nowak, Kristine L., and Christian Rauh. "The Influence of the Avatar on Online Perceptions of Anthropomorphism, Androgyny, Credibility, Homophily, and Attraction." *Journal of Computer-Mediated Communication* 11.1 (2005): Article 8. Web. 29 Nov. 2009.

O'Brien, Tim. *The Things They Carried.* New York: Broadway Books, 1990. Print.

Olson, Lester C., Cara A.Finnegan, and Diane S. Hope. "Visual Rhetoric in Communication: Continuing Questions and Contemporary Issues." *Visual Rhetoric: A Reader in Communication and American Culture.* Ed. Lester C. Olson, Cara A. Finnegan, and Diane S. Hope. Thousand Oaks: Sage, 2008. 1–14. Print.

Parks, Lisa. "Plotting the Personal: Global Positioning Satellites and Interactive Media." *Ecumene: A Journal of Cultural Geographies* 9.2 (2001): 209–222. *Academic Search Premiere.* Web. Sept. 2008.

Park Ranger. Personal Interview. Lowell Mills National Historical Park. Lowell, MA. 10 March 2007.

Pauly, D., V. Christensen, R. Froese, A. Longhurst, T. Platt, S. Sathyendranath, K. Sherman, and R. Watson. "Mapping Fisheries onto Marine Ecosystems: A Proposal for a Consensus Approach for Regional, Oceanic, and Global Intersections." *Methods for Evaluating the Impacts of Fisheries on North Atlantic Ecosystems.* Ed. Daniel Pauly and Tony Pitcher. Fisheries Centre Research Reports 8.2. Vancouver: University of British Columbia, 2000. Web. January 2010.

Pickles, John. Foreword. *The Natures of Maps: Cartographic Constructions of the Natural World.* By Denis Wood and John Fels. Chicago: U of Chicago P, 2008. Print.

—. *A History of Spaces: Cartographic Reason, Mapping and the Geo-Coded World.* New York: Routledge, 2004. Print.

—. "Representations in an Electronic Age: Geography, GIS, and Democracy." *Ground Truth: The Social Implications of Geographic Information Systems.* Ed. John Pickles. New York: Guildford, 1995. 1–30. Print.

Plato. *Phaedrus.* Trans. Alexander Nehamas and Paul Woodruff. Cambridge: Hackett, 1995. Print.

Prelli, Lawrence J. "Rhetorics of Display: An Introduction." *Rhetorics of Display.* Ed. Lawrence J. Prelli. Columbia: U of South Carolina P, 2006. 1–40. Print.

—. "Visualizing a Bounded Sea: A Case Study in Rhetorical Taxis." *Rhetorics of Display.* Ed. Lawrence J. Prelli. Columbia: U of South Carolina P, 2006. 90–120. Print.

Propen, Amy D. "Cartographic Representation and the Construction of Lived Worlds: Understanding Cartographic Practice as Embodied Knowledge." *Rethinking Maps: New Frontiers in Cartographic Theory.* Ed.

Martin Dodge, Rob Kitchin, and Chris Perkins. New York: Routledge Studies in Human Geography, 2009. 113–130. Print.

—. "Critical GPS: Toward a New Politics of Location." *ACME: An International E-Journal of Critical Geographies. Special Issue: Critical Cartographies* 4.1 (2005): 131–144. Print.

—. "Visual Communication and the Map: How Maps as Visual Objects Convey Meaning in Specific Contexts." *Technical Communication Quarterly* 16 (2007): 233–254. Print.

Propen, Amy, and Mary Lay Schuster. "Making Academic Work Advocacy Work: Technologies of Power in the Public Arena." *Journal of Business and Technical Communication* 22 (2008): 299–329. Print.

—. "Understanding Genre through the Lens of Advocacy: The Rhetorical Work of the Victim Impact Statement." *Written Communication* 27.1 (2010): 3–35. Print.

Reynolds, Nedra. *Geographies of Writing: Inhabiting Places and Encountering Difference.* Carbondale: Southern Illinois UP, 2004. Print.

Richards, Anne R., and Carol David. "Fields of Vision: A Background Study of References for Teachers." *Writing the Visual: A Practical Guide for Teachers of Composition and Communication.* Ed. Carol David and Anne R. Richards. West Lafayette: Parlor Press, 2008. 3–31. Print.

Rogoff, Irit. "Studying Visual Culture." *Visual Rhetoric in a Digital World: A Critical Sourcebook.* Ed. Carolyn Handa. New York: Bedford/St. Martin's, 2004. 381–394. Print.

Rose, Gillian. *Visual Methodologies: An Introduction to the Interpretation of Visual Materials.* Thousand Oaks, CA: Sage, 2001. Print.

Rothenberg, Ellen. *Circular Fence Sculpture with Poem.* 1995. Metal Sculpture. Lowell Mills Natl. Historical Park, Lowell, MA.

—. *Fourteen-Hour Clock.* 1995. Granite Sculpture. Lowell Mills National Historical Park, Lowell, MA.

—. *Path Markers.* 1995. Sculpture. Lowell Mills Natl. Historical Park, Lowell, MA.

—. *Seating Circle: Truth Loses Nothing Upon Investigation.* 1995. Granite Sculpture. Lowell Mills National Historical Park. Lowell, MA.

—. *Steps.* 1995. Granite Sculpture. Lowell Mills Natl. Historical Park, Lowell, MA.

"Sarah George Bagley." *Center for Lowell History.* U of Massachusetts Lowell Libraries. N.d. Web. March 2007.

Schuster, Mary Lay. "A Different Place to Birth: A Material Rhetoric Analysis of Baby Haven, a Free-Standing Birth Center." *Women's Studies in Communication* 29 (2006): 1–38. Print.

Schuurman, Nadine. "Trouble in the Heartland: GIS and Its Critics in the 1990s." *Progress in Human Geography* 24 (2000): 569–590. Print.

Schuurman, Nadine, and Geraldine Pratt. "Care of the Subject: Feminism and Critiques of GIS." *Gender, Place and Culture* 9 (2000): 291–299. Print.

Segal, Judy. *Health and the Rhetoric of Medicine.* Carbondale: U of Southern Illinois P, 2005.

Shipka, Jody. "Sound Engineering: Toward a Theory of Multimodal Soundness." *Computers and Composition* 23 (2006): 355–373. *Academic Search Premiere.* Web. Aug. 2008.

Siebert, Charles. "Watching Whales Watching Us." *The New York Times.* 8 July 2009. Web. 10 July 2009.

Spinuzzi, Clay. "Pseudotransactionality, Activity Theory, and Professional Writing Instruction." *Technical Communication Quarterly* 5 (1996): 295–308. Print.

—. *Tracing Genres Through Organizations: A Sociocultural Approach to Information Design.* Cambridge: MIT Press, 2003.

Stanton, Cathy. *The Lowell Experiment: Public History in a Postindustrial City.* Amherst: U of Massachusetts P, 2006. Print.

Strauss, Anselm, and Juliet Corbin. "Grounded Theory Methodology: An Overview." *Strategies of Qualitative Inquiry.* Ed. Norman K. Denzin and Yvonna S. Lincoln. Thousand Oaks: Sage, 1998. 158–183. Print.

Sturken, Marita, and Lisa Cartwright. *Practices of Looking: An Introduction to Visual Culture.* Oxford: Oxford UP, 2001. Print.

Sullivan, Patricia Suzanne, et al. *Writing on the Line: Rhetoric in Boston's Southwest Corridor.* Northeastern University, 2008. Web. January 2010.

Tufte, Edward. *The Visual Display of Quantitative Information.* Cheshire: Graphics P, 1983. Print.

Turnbull, David. *Maps Are Territories, Science Is an Atlas: A Portfolio of Exhibits.* Chicago: U of Chicago P, 1989. Print.

United States. Dept. of Commerce. Natl. Oceanic and Atmospheric Administration. "Taking and Importing Marine Mammals; Taking Marine Mammals Incidental to Navy Operations of Surveillance Towed Array Sensor System Low Frequency Active Sonar; Final Rule." 50 CFR Part 216. *Federal Register* 67.136 (16 July 2002): 46712–46789. Web. Jan. 2010.

University of Minnesota Press. "Overview of *What Is Posthumanism?* by Cary Wolfe." June 2009. Web. December 2009.

Waddell, Craig. "Saving the Great Lakes: Public Participation in Environmental Policy." *Green Culture: Environmental Rhetoric in Contemporary America.* Ed. Carl G. Herndl and Stuart C. Brown. Madison: U of Wisconsin P, 1996. 141–165. Print.

Warnick, Barbara. "Looking to the Future: Electronic Texts and the Deepening Interface." *Technical Communication Quarterly,* 14 (2005): 327–333. Print.

Weeks, Linton. "From Maps to Apps: Where are We Headed?" *National Public Radio.* 4 May 2010. Web. May 2010.

Wolfe, Cary. *Animal Rites: American Culture, the Discourse of Species, and Posthumanist Theory.* Chicago: U of Chicago P, 2003. Print.

Wood, Denis. *The Power of Maps.* New York: Guilford, 1992. Print.

Wood, Denis, and John Fels. "Designs on Signs: Myth and Meaning in Maps. *Cartographica* 23 (1986): 54–103. Print.

—. "Don't Skip This." Introduction. *The Natures of Maps: Cartographic Constructions of the Natural World.* By Wood and Fels. Chicago: U of Chicago P, 2008. xv-xvi. Print.

Wortham, Jenna. "Sending GPS Devices the Way of the Tape Deck?" *The New York Times.* 7 July 2009. Web. 8 July 2009.

Wright, Elizabethada A. "Rhetorical Spaces in Memorial Places: The Cemetery as a Rhetorical Memory Place/Space." *Rhetoric Society Quarterly* 35.4 (2005): 51–81. Print.

Xia Jianhong (Cecilia), Colin Arrowsmith, Mervyn Jackson, and William Cartwright. "The Wayfinding Process Relationships Between Decision-making and Landmark Utility." *Tourism Management* 29 (2008): 445–457. Print.

Index

About the Author

Amy D. Propen is Assistant Professor of Rhetoric and Composition at York College of Pennsylvania, where she teaches courses in the Professional Writing program. She received her PhD in Rhetoric and Scientific and Technical Communication from the University of Minnesota. Her research on visual rhetoric, critical cartographies, and rhetoric as advocacy has appeared in *Technical Communication Quarterly*; *Journal of Business and Technical Communication*; *Written Communication*; *Law, Culture and the Humanities*; *ACME: An International E-Journal of Critical Geographies*; and the edited collection, *Rethinking Maps: New Frontiers in Cartographic Theory*. She is co-author, with Mary Lay Schuster, of *Victim Advocacy in the Courtroom: Persuasive Practices in Domestic Violence and Child Protection Cases*.

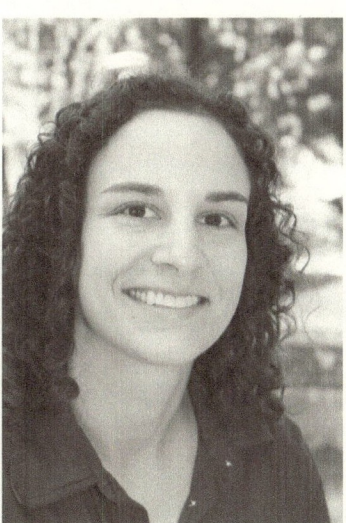

Photograph of the author by Michael Daniel Adams, York College of Pennsylvania. Used by permission.